BASIC PHYSICS OF ULTRASONOGRAPHIC IMAGING

Diagnostic Imaging and Laboratory Technology
Essential Health Technologies
Health Technology and Pharmaceuticals
WORLD HEALTH ORGANIZATION
Geneva

BASIC PHYSICS OF ULTRASONOGRAPHIC IMAGING

Editor
Harald Ostensen

Author
Nimrod M. Tole, Ph.D.
Associate Professor of Medical Physics
Department of Diagnostic Radiology
University of Nairobi

WORLD HEALTH ORGANIZATION

WHO Library Cataloguing-in-Publication Data

Tole, Nimrod M.
  Basic physics of ultrasonic imaging / by Nimrod M. Tole.

  1.Ultrasonography I.Title.

  ISBN 92 4 159299 0   (NLM classification: WN 208)

© World Health Organization 2005

All rights reserved. Publications of the World Health Organization can be obtained from WHO Press, World Health Organization, 20 Avenue Appia, 1211 Geneva 27, Switzerland (tel: +41 22 791 2476; fax: +41 22 791 4857; email: bookorders@who.int). Requests for permission to reproduce or translate WHO publications – whether for sale or for noncommercial distribution – should be addressed to WHO Press, at the above address (fax: +41 22 791 4806; email: permissions@who.int).

The designations employed and the presentation of the material in this publication do not imply the expression of any opinion whatsoever on the part of the World Health Organization concerning the legal status of any country, territory, city or area or of its authorities, or concerning the delimitation of its frontiers or boundaries. Dotted lines on maps represent approximate border lines for which there may not yet be full agreement.

The mention of specific companies or of certain manufacturers' products does not imply that they are endorsed or recommended by the World Health Organization in preference to others of a similar nature that are not mentioned. Errors and omissions excepted, the names of proprietary products are distinguished by initial capital letters.

All reasonable precautions have been taken by WHO to verify the information contained in this publication. However, the published material is being distributed without warranty of any kind, either express or implied. The responsibility for the interpretation and use of the material lies with the reader. In no event shall the World Health Organization be liable for damages arising from its use.

The named author alone is responsible for the views expressed in this publication.

Printed in Malta, by Interprint Limited.

Acknowledgement

I wish to thank Dr. Harald Ostensen of the World Health Organization for inspiring me to embark on the task of writing this book. I am most grateful to Miss Susan Mwanzia for her dedicated and skilful work during the preparation of the manuscript.

PREFACE

Ultrasound as a tool for medical imaging evolved rapidly during the second half of the 20th century. Its value became established in those areas of imaging in which its unique features offer special advantages over other imaging modalities. It gained wide acceptance as a non ionizing form of energy which is applied non-invasively. The demand for basic as well as specialized ultrasound services has grown steadily, and continues to grow, with new horizons still being explored.

The development and expansion of basic ultrasound services is more readily achievable in comparison to other imaging technologies because it is less demanding in terms of the physical infrastructure required, and the costs of equipment and consumables. However, the same cannot be said in reference to personnel. The provision of reliable diagnostic ultrasound services requires well trained, highly skilled, resource persons. The training of personnel must therefore be regarded as a major responsibility on the part of those providing ultrasound services, whether basic or advanced. Some efforts have been made to prescribe the training requirements and qualifications of those wishing to practise diagnostic ultrasound. However, there has been no consensus on the subject. Some individual countries and professional organizations do provide regulatory guidelines, but generally the situation remains somewhat amorphous, and perhaps unsatisfactory. The difficulties in addressing this issue arise mainly from the wide diversity in the cadres of personnel who may be allowed to practice in different countries. In the developing countries, where usage of diagnostic ultrasound has been increasing quite rapidly, formal training courses and certification have been scarce.

Whereas the benefits of clinical ultrasound applied properly are widely acknowledged, it must be pointed out, indeed emphasized, that when the technology is left in the hands of the untrained, it impacts negatively on patient care.

Therefore the quest to promote ultrasound usage and make it more accessible to wider sectors of humanity must be accompanied by concerted efforts to train the personnel charged with the responsibility of providing the services to patients. This manual is intended to make a humble contribution towards these efforts in education and training.

Training in medical ultrasound is, of necessity, a multidisciplinary exercise. Besides medical knowledge and a good background in anatomy, one also needs to be well exposed to the physical principles underlying the use of ultrasound in imaging, and to the practical techniques of applying the technology. This text is dedicated to the basic physics of ultrasonic imaging.

Knowledge of the basic physics of ultrasound is essential as a foundation for the understanding of the nature and behaviour of ultrasound, the mechanisms by which it interacts with matter, the process of image formation, the choice of imaging parameters, the optimization of image characteristics, and the identification of artefacts. These topics are highlighted in this book. A good foundation in the fundamentals of imaging physics prepares practitioners to use their equipment more optimally in order to generate high quality images, and to make more critical evaluations of the images they produce. In turn, good practice leads to improved patient care.

Granted that some knowledge of physics is desirable, an important question arises: how much physics is the ultrasound practitioner expected to learn? The answers will vary, bearing in mind the variety of professional groups which may be considered as a pool for recruitment. Some of the professional cadres may have had only a rudimentary background in the physical sciences, while others may be regarding physics as a difficult and awesome subject. Therefore, while the need to impart some knowledge on the physical basis of ultrasonic imaging to all potential practitioners must be emphasized, the limitations among many of those targeted should also be acknowledged and appropriately addressed.

These considerations suggest that there is a need for simple educational material on the physics of ultrasonic imaging. This manual has been prepared in response to that need. It attempts to present the subject from a simple approach that should make it possible for the target groups to comprehend the important concepts which form the physical basis of ultrasonic imaging. The main target group of this manual is radiological technologists and radiographers working with diagnostic ultrasound in developing countries. Clinicians and nurse practitioners may also find the simple presentation appealing. A conscious effort has been made to avoid detailed mathematical treatment of the subject. The emphasis is on simplicity. All in all, if the average radiological technologist or radiographer finds the text "readable", then this goal will have been achieved.

It is not lost to the author that a zealous attempt to simplify scientific concepts is fraught with the risks of compromising on scientific accuracy. Every effort has been made to avoid such pitfalls. Nevertheless, the author accepts responsibility for any factual errors that may have inadvertently escaped his attention.

INTRODUCTION

As in other areas of medical imaging, physics plays a leading role as the foundation for ultrasonic imaging. A brief insight into the physical concepts involved will be outlined here, while more detailed treatment of specific subject areas will be presented in the main text of the manual.

Ultrasonic imaging is a technique of generating images using a very high frequency sound. Sound is a mechanical, vibration form of energy. Ultrasound for medical imaging is generated in special crystalline materials which, when electrically excited, are capable of vibrating at frequencies of millions of vibrations per second. The devices in which ultrasound is produced, and also detected, are called transducers.

Once produced, the ultrasonic energy is suitably focused into a narrow beam, which is then directed into the tissues in selected areas of interest in the patient. Along its path, the beam interacts with the tissues through various processes, suffering a reduction in its intensity, or attenuation. The major interaction processes include reflection, refraction, absorption, and scattering of the beam energy. The beam also undergoes changes in shape and size as it spreads out, or diverges, beyond the region where it is well-focused. The interaction processes are affected by both the parameters of the ultrasound beam, especially the frequency, and the physical properties of the medium through which the beam passes. The most significant medium properties in this respect include the density, elasticity, and viscosity. The behaviour of some biological materials is worthy of note. The presence of gas is associated with near-total reflection of ultrasound. On the other hand, liquid compartments do not reflect and therefore present as characteristic echo-free zones. Bone is noted for its heavy absorption. The soft tissues show comparatively less dramatic behaviour under interrogation by ultrasound.

The attenuation processes have an influence on the generation of the ultrasonic image. That part of the beam energy which is reflected or scattered backwards in the direction of the transducer provides information concerning the structures along the beam path. Reflection occurs at tissue interfaces, otherwise referred to as acoustic boundaries. The reflected echoes are detected by the transducer and converted into electrical signals, which are then measured and displayed on a cathode ray oscilloscope. The sizes of the echoes and their locations are determined, and these provide the useful diagnostic parameters from which the ultrasonic image is built. The resulting echo pattern will be characteristic of the different tissue structures lying within the beam path.

The attenuation processes have the effect of reducing the intensity of the ultrasound beam as it traverses the tissues. The intensity varies not only with the type of medium, but also with distance within a specified medium. The impact of this reduction on the imaging process is that, at some tissue depth, the remaining intensity will become too low to provide echoes that are large enough to be useful. A strongly reflecting interface such as that between gas and soft tissue will send back almost all the energy beamed. It acts as a barrier to the propagation of ultrasound. Similarly, a strongly absorbing medium such as bone removes energy from the beam by converting the energy it absorbs from the beam into another form of energy, usually heat. Such a medium then casts an acoustic shadow behind it, because too little of the beam energy is left for further interrogation of the structures lying beyond it. In contrast, liquid cavities are an invaluable asset to ultrasonic imaging, because they do not cause reflection or refraction of ultrasound. Consequently, besides showing their characteristic echo-free appearance, they also serve as acoustic windows through which ultrasound travels long distances.

Another consequence of beam attenuation which affects image formation relates to the sizes of echoes. Whereas similar reflecting interfaces are expected to give rise to equal-sized echoes, this will not be the case if the interfaces are located at different tissue depths. In practical imaging situations, this complication is addressed through electronic compensation of echo sizes, a process called Time-Gain Compensation.

The intensity of an ultrasound beam, and the related quantity called ultrasonic power, are important in two ways. First, the output intensity of an ultrasound instrument affects its sensitivity, and secondly, knowledge of these quantities is required when we wish to evaluate the potential biological consequences of exposure to ultrasound. The intensity of an ultrasound beam can be specified either in absolute terms or using a relative scale.

The absolute measure is stated as the rate at which energy flows through unit cross-sectional area at a point in the propagating medium, and quantified in units such as joule per second per square meter. On the relative scale, a suitable reference intensity is identified and all other intensities are then compared to this reference. Relative intensities are expressed in decibels (dB). For example, the output intensity of an ultrasound instrument could be chosen as the reference intensity and assigned the decibel value 0 dB. Intensities at different tissue depths would then have negative decibel values due to attenuation. Each of the two methods of specifying intensity has its own merits. If we merely wish to express the intensity of the beam at some point in tissue as a ratio of the surface intensity, the dB scale would be quite convenient. However, if we sought to establish the total energy deposited in a volume of tissue for purposes of estimating the potential biological consequences, then we would need to resort to absolute values of intensity and exposure duration.

During diagnostic applications, ultrasound can be beamed either continuously, or in the form of short, intermittent pulses, with intervals between bursts. The two methods of application are referred to as continuous wave (CW) ultrasound and pulsed wave (PW) ultrasound, respectively. Most of diagnostic ultrasound employs what is known as the pulse-echo principle. The transducer is pulsed very briefly to send a burst of ultrasonic energy into the tissues. This is then followed by a much longer "listening" phase when the transducer does not emit but receives and detects returning echoes. This emit-detect sequence is then repeated at a high rate called the pulse repetition frequency. The same transducer can be used to produce the beam and to detect the echoes. The equipment measures the time intervals between the emission of a pulse and the detection of echoes at the transducer, and, using the known velocity of ultrasound in soft tissue, calculates the distances of the various reflectors from the transducer. In CW ultrasound, emission is continuous, and a separate detector is required to receive the returning echoes. This mode is employed to compare the frequencies of the transmitted beam and that of the echoes received from moving structures to provide useful information during flow studies. The basis of this technique is known as the Doppler effect. Whereas PW techniques facilitate distance measurements, this cannot be achieved with CW ultrasound alone. However, in advanced techniques of diagnostic ultrasound, a combination of PW and CW ultrasound are employed to generate images during flow studies.

The data acquired from returning echoes is used to determine the magnitude of each echo received, and the distance of its associated reflector from the transducer position. This information is then integrated to build up the ultrasound image, or to display motion of structures. The image is then displayed on a cathode ray oscilloscope, or a TV monitor, from which a hard copy may be captured, for example using thermal printing paper. Different methods are used to record and display the acquired information. These display modes are variously described as A-mode, B-mode, M-mode, and so on. Some of them provide information in a single dimension, others in 2-dimension, and yet others are specifically designed to reveal moving structures.

In the early days of ultrasonic imaging, only static images could be generated, by moving the ultrasound transducer manually over the region of interest. Due to the slow speed of acquisition, the static image did not provide the true picture of the scan section at any given moment. In addition, relating echo information to the corresponding geometrical coordinates in the subject (i.e. determining the points of origin of the echoes) was a complicated matter. Advancements in electronics, transducer design, and computer technology have led to evolution in ultrasonic imaging that now facilitates the generation and display of dynamic, real-time images. Real-time imaging enables the simultaneous observance of moving structures over the whole image plane. Further advances continue to be realized.

Image quality is a vital consideration in any imaging system. A variety of factors contribute to the overall quality of the ultrasound image. These include the design of equipment components, especially the transducer, the choice of imaging parameters, particularly the beam frequency, and the skilful use of the equipment by the operator. A good quality image should contain information that is associated with high spatial resolution (ability to distinguish between objects in space), high contrast resolution (ability to distinguish between signals of different size), and high temporal resolution (ability to separate between events in time). In addition, the image should be free of any avoidable artefacts. The enhancement of a particular image characteristic is dependent upon the appropriate selection of imaging parameters. There is an intimate interplay between image characteristics in ultrasonic imaging, in which efforts to enhance a particular characteristic often diminishes another desirable characteristic. For example, the selection of a high beam frequency enhances spatial resolution but leads to increased attenuation of the beam, hence reduced tissue depth. Many such compromise situations necessitate a give-and-take approach in the choice of the selectable imaging parameters. It is essential for the ultrasound practitioner to have a good grasp of the various opposing interests, since appropriate decisions must be made on how to optimize the desired image characteristics.

In the routine process of generating ultrasonic images, some information that does not represent the true picture may be recorded. Such information may obscure otherwise true and useful information, or mislead the practitioner into arriving at the wrong conclusions. There are many possible causes for such artefacts. Some of them are avoidable, whereas others are an inherent feature of the imaging technique. Due attention must be paid to the possible presence of artefacts on images.

Whenever any form of energy is used to irradiate humans in medical procedures, consideration must be given to the safety of its usage. The potential biological consequences of exposure, and the relationship between dose and effect, must be studied. The use of ultrasound has been subjected to extensive scrutiny in this respect. Laboratory studies have established that the potential biological effects arise secondary to the elevation of tissue temperature, and to the formation and collapse of bubbles in liquids. Most fortunately, the intensity levels and exposure durations at which those effects have their thresholds are hundreds of times those employed in diagnostic ultrasound. This makes ultrasound one of the safest forms of energy used in medical imaging.

This overview of the physical concepts underlying the use of ultrasound should serve to indicate the strong linkage between ultrasonic imaging and the science of physics, and to emphasize the need for practitioners of ultrasound to acquire some basic knowledge on these concepts. In the chapters that follow, the physical principles of ultrasonic imaging will be presented in greater detail.

CHAPTER 1

THE NATURE OF ULTRASOUND

1.1 The sound spectrum

Sound is a mechanical form of energy. A vibrating source is responsible for the production of sound. The number of vibrations per unit time, called the **frequency** of vibrations, determines the quality of the sound produced. Frequency is expressed in units called **hertz**, abbreviated Hz.

$$1 \text{ Hz} = 1 \text{ vibration per second.}$$

The sound spectrum can be conveniently divided into three distinct parts. **Audible sounds** are those which can be perceived by the human ear. There are some differences between individuals in their ability to perceive sound frequencies. In most humans, the audible frequency range is approximately 20 Hz - 20,000 Hz. Sound which has a frequency below that which can be perceived by the human ear is referred to as **infrasound**, while sound of frequencies higher than that of human perception is known as **ultrasound.** Therefore, ultrasound may be defined as sound energy of frequency higher than 20 kilohertz (20 kHz).

1.2 Propagation of ultrasound

1.2.1 Transfer of energy

The propagation of ultrasonic energy requires a material medium: it cannot take place in empty space. A source of ultrasound in contact with a medium transfers the mechanical disturbance to the medium, initiating vibrations in the "particles" of the medium. The term "particle" as used here is not uniquely defined. It is often equated to the molecules of the material medium. A particle may be conceptualised as representing a very small volume of matter within the medium, in which all the atoms contained in that volume respond uniformly to a physical stimulation. A vibrating particle performs microscopic to-and-fro movement about its mean position within the medium. Through these tiny vibration movements, neighbouring particles are similarly affected, setting them in motion by **direct transfer of energy** from one particle to another. As each vibrating particle transfers energy to its immediate neighbours, the bulk of the medium is quickly pervaded by the mechanical vibrations. In effect, ultrasonic energy is propagated through the medium. It is important to understand that propagation of the ultrasonic energy does not entail actual migration of particles across the medium: the particles merely oscillate about their mean positions.

1.2.2 Pressure waves

The mechanical movements of a source of ultrasound may be compared to the action of a piston moving rapidly in confined space. In the forward direction, the piston compresses the medium particles in front of it, increasing their concentration per unit volume, hence creating increased pressure. This is referred to as the **compression phase**, sometimes also called the **condensation phase**. When the piston moves in the reverse direction, the medium particles are decompressed, giving rise to a low pressure phase, known as **rarefaction.** The periodic movement of the piston therefore creates pressure waves in front of it, alternating between high and low pressure.

Similarly, the mechanical vibrations of a source of ultrasound create alternating phases of compression and rarefaction in the particles of the propagating medium with which it is in contact. These pressure variations are propagated through the transmitting medium. Figure 1.1. illustrates these variations. Closely packed lines represent the compression phase, while regions of low line density represent rarefaction.

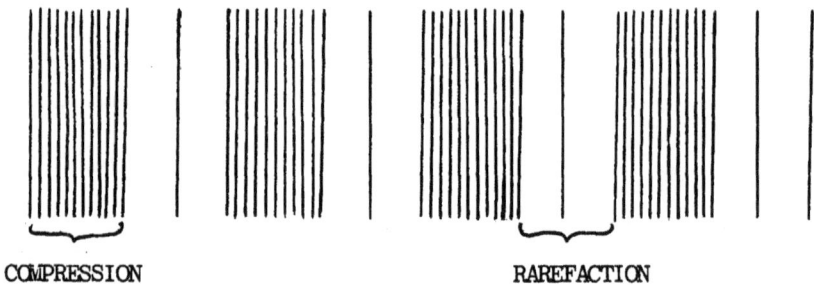

Fig. 1.1 Pressure variations in the propagation of ultrasound

1.2.3 Longitudinal propagation

In the propagation of ultrasound, the direction of displacement of medium particles is usually the same as the direction of oscillation of the source of ultrasound. Therefore, we can say that the ultrasound wave is propagated in the same direction as that of the disturbance causing it. Such waves are called **longitudinal waves** (see Fig. 1.2).

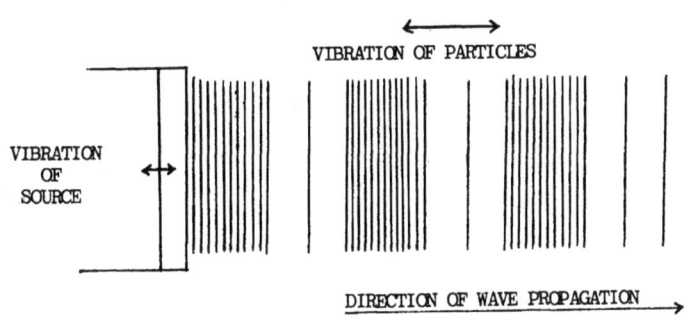

Fig. 1.2 Longitudinal propagation of ultrasound

When propagation of a wave takes place in a direction that is perpendicular to that of the disturbance causing it, then the wave is referred to as a **transverse wave.** Although ultrasound is usually propagated as longitudinal waves, it should be noted that in bones it may be propagated as transverse waves.

1.2.4. Simple wave parameters.

The periodic movements of medium particles about their mean positions, and the corresponding regular fluctuations in pressure, can be conveniently represented as a sinusoidal curve. (Fig 1.3)

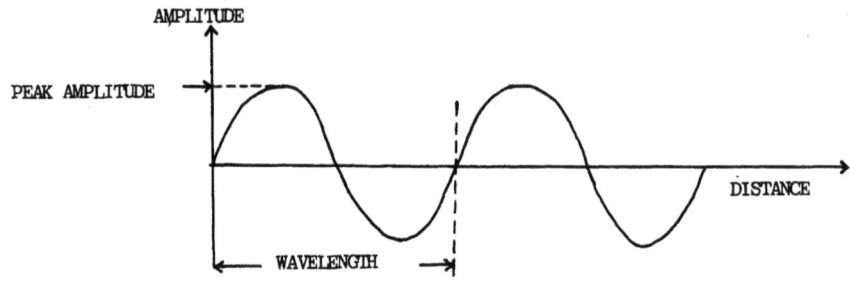

Fig 1.3 Variation of particle displacement in the direction of wave travel represented as a sinusoidal curve.

The sine waveform is characterized by the wave parameters of **amplitude** and **wavelength.** The amplitude at a given moment is the magnitude of particle displacement from its mean position at that particular time. The **peak amplitude** represents the maximum particle displacement. It corresponds to the maximum change in pressure, and represents the vigour of vibrations, which, in turn, is related to the intensity of the ultrasound beam.

When the amplitude is plotted against distance, the **wavelength** is the distance between two consecutive, corresponding positions on the sine wave. It represents the distance travelled by the pressure wave during one complete wave cycle. Each complete wave cycle is attributed to one vibration of the source. The number of vibrations per unit time is called the **frequency** of vibrations. The frequency also represents the number of times the wave is repeated per unit time. Therefore, the product of wavelength and frequency gives the total distance travelled by the wave in unit time - in other words, the **velocity of the wave**.

Eg 1.1: Wavelength x frequency = wave velocity

Ultrasound for medical imaging employs frequencies in the megahertz (MHz) range.

1 MHz = 1 million vibrations per second.

If in Fig 1.3 the amplitude were plotted against time, the **period** of the wave would be shown as the time for one complete wave cycle. The wave period is equal to the reciprocal of the frequency.

Only simple wave parameters describing continuous waves have been introduced in this section. Other wave parameters relevant to pulsed waves are mentioned elsewhere in the manual.

1.2.5 Velocity of ultrasound

The rate at which the ultrasound wave is propagated through a medium is called the wave velocity. This velocity varies from one medium to another, depending on the elastic properties of the material. Two physical properties of the medium are crucial in this respect. These are the **density** and the **compressibility** of the medium.

Density affects the mass content of medium particles. Denser materials tend to be composed of more massive particles. The propagation of the ultrasound wave is associated with periodic movement of medium particles. When these particles are massive, greater force is required to initiate particle movement, and also to bring to a halt such a moving particle. In other words, the more massive a particle is, the greater its inertia. The larger force required to overcome particle inertia in denser materials leads to the conclusion that **wave velocity should be lower in materials of high density**. If this conclusion appears to contradict common experience, it is because of the role of the second factor, the compressibility.

Compressibility refers to the ease with which a medium can be mechanically deformed and reformed. Other terms commonly used to describe this property are **stiffness** and **rigidity**. There is a relationship between compressibility and density. Materials of low density such as gas are easier to compress, because their particles are farther apart. They have **high** compressibility. In contrast, high density materials have low compressibility.

The effect of compressibility on the velocity of ultrasound may be understood by considering the extent of particle movement required to facilitate wave propagation. In a material of low compressibility, such as bone, the closeness of particles to one another implies that only slight movement of particles is required in order to effect transfer of energy to their neighbours - transfer of energy is more rapid. **This predicts a higher wave velocity in materials of low compressibility.**

The wave velocity as observed in practice represents the combined effects of medium density and compressibility. One factor may play the more predominant role, or the effects of the two factors may moderate each other. Table 1 shows the velocities of ultrasound in air, soft tissue (average), and bone. Among these materials, the velocity is highest in bone, and lowest in air. While the higher density of bone predicts reduced velocity of ultrasound, its lower compressibility increases the velocity to a greater extent. Compressibility is the more predominant factor here.

TABLE 1: VELOCITY OF ULTRASOUND IN AIR, BONE, AND SOFT TISSUE.

Material	**Velocity (m/s)**
Air (NTP)	330
Bone	3,500 - 4,800
Soft tissue	1,540 (average)

Two other factors which in theory affect the velocity of ultrasound should be mentioned. There is a slight variation of the velocity with beam frequency (a phenomenon called **dispersion**). However, in the whole of the diagnostic frequency range from 1 MHz to 20 MHz, the variation is less than 1%, so this effect can be considered as insignificant. The velocity also changes with medium temperature, but for a few degrees temperature shift, the velocity change is quite small. Since body temperature is under homeostatic control, the effect of temperature on velocity is negligible in clinical ultrasound.

The velocities of ultrasound in the various soft tissues are quite identical. The average value of 1,540 m/s is worthy of note. First, this value is used in the calibration of distance measurements in diagnostic ultrasound. Secondly, it has a very fundamental significance to the practice of ultrasonic imaging. On the one hand, this velocity turns out to be a most fortuitous blessing in that it is sufficiently high to facilitate the collection of imaging data at typical tissue distances within the short durations desirable for the purpose. It takes approximately 65 microseconds for ultrasound to travel a distance of 10 cm in soft tissue (or to receive an echo from a tissue depth of 5 cm). If the velocity of ultrasound in soft tissue were substantially lower, the scope of clinical ultrasound would have faced major inherent constraints. On the other hand, most of the trade-offs that must be made between image quality, tissue depth and framing rates during real-time imaging can be attributed to the soft tissue velocity not being high enough for the ideal combinations! This subject is discussed further elsewhere in the text.

CHAPTER 2

GENERATION AND DETECTION OF ULTRASOUND

2.1 The piezoelectric phenomenon

Although a number of different methods are available for the production of high frequency mechanical vibrations, the most common method of generating ultrasound, and the one employed in the devices used in clinical ultrasound, relies on a phenomenon called the piezoelectric effect. This phenomenon is exhibited by some crystalline materials, and involves the reversible conversion of two forms of energy from one to the other, namely mechanical and electrical energies.

The prefix piezo means pressure. When crystals of piezoelectric materials are compressed or stretched (i.e., when mechanical stress is applied upon them), electrical charge will appear on their surfaces. Mechanical energy will have been transformed into electrical energy. This process is called the **piezoelectric effect**. Figure 2.1 illustrates this effect, and also shows that the polarity of the induced surface charge is reversed between compression and stretching.

Fig. 2.1   The piezoelectric phenomenon

Conversely, when a potential difference is applied between the faces of a piezoelectric crystal, the crystal will respond by expanding or contracting. Electrical energy will have been converted to mechanical energy. This is the **reverse piezoelectric effect.**

2.2 Production and detection of ultrasound

Recalling that ultrasound is a mechanical form of energy, we can deduce that to produce ultrasound using the piezoelectric phenomenon we must rely on that process which converts electrical energy into mechanical energy, i.e. the reverse piezoelectric effect. On the other hand, we can use the same phenomenon to detect high frequency mechanical vibrations, by converting them into an electrical signal (the piezoelectric effect). In summary, the generation and detection of ultrasound is done by using crystals of piezoelectric materials. **Production of ultrasound relies on the reverse piezoelectric effect, while detection is based on the piezoelectric effect.** Because of the reversibility of this phenomenon, it is possible to use the same crystal to produce ultrasound, and subsequently to detect echoes returning to the crystal from a reflector at some distance away.

2.3 Ultrasonic transducers

Devices which convert one form of energy into another are called transducers. The probe assembly used to generate and detect ultrasound is therefore aptly called a transducer. There are several different designs of ultrasonic transducers. First, we consider a transducer made with a single piezoelectric crystal. Multi-crystal transducers will be discussed in Chapter 7.

2.3.1 The single crystal transducer

The essential components of the single crystal transducer are shown in Fig 2.2

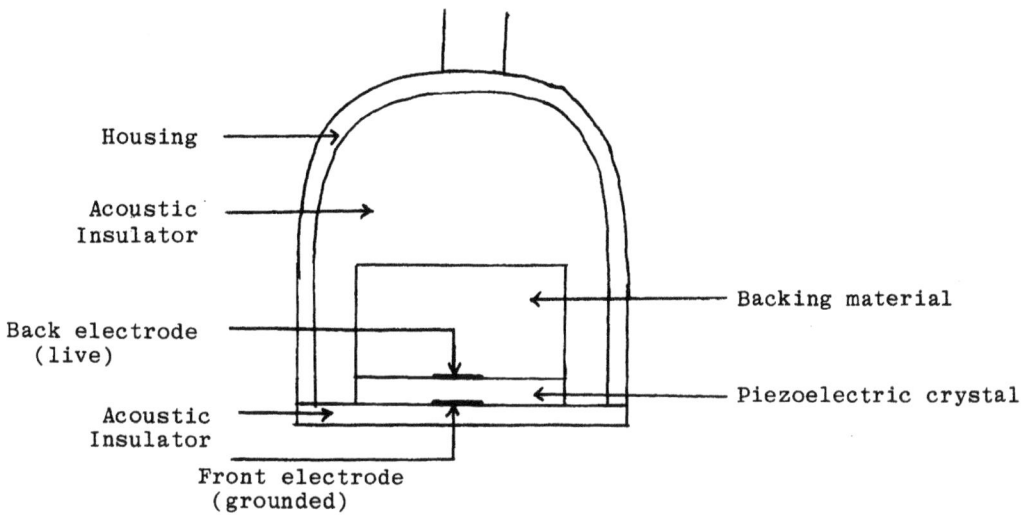

Fig 2.2   Components of a single crystal (T)

The crystal element

The crystal element is the most important component of the transducer. It is a thin disc of piezoelectric material near the front surface of the transducer. The crystal material may possess its piezoelectric properties naturally, but more commonly, the piezoelectric properties are artificially induced using a combination of thermal and electrical treatment. It is important to note that the piezoelectric properties of an artificial crystal can be destroyed if the crystal is heated to high temperatures. For this reason, **the sterilization of ultrasonic transducers should not be done by autoclaving.**

The crystal thickness controls the frequency of vibrations. A vibrating crystal will transmit ultrasound in both directions from its two surfaces. The crystal thickness is chosen such that the vibrations at the two surfaces will reinforce each other every time the ultrasound makes a round trip internally from one face to the other and back to the first face. Reinforcement takes place if the distance covered by the ultrasound during the round trip equals a whole wavelength of the ultrasound wave. The wave then arrives at a crystal face in exactly the same phase of the wave cycle as the preceding disturbance which caused it. Consecutive vibrations on the crystal faces will then reinforce each other through what is known as **constructive interference** in wave theory. The reinforcements result in prolonged self-sustenance of vibrations, a condition called **resonance.**

Fig 2.3 shows that conditions of resonance are met when the thickness, t, of the piezoelectric crystal is equal to one half of the wavelength corresponding to the desired frequency of vibrations.

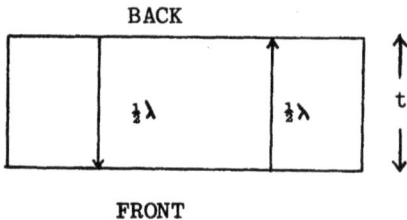

Fig. 2.3 Thickness of a crystal determines vibrational frequency

The relationships used to determine the thickness corresponding to a particular frequency are:

(i) frequency = $\dfrac{\text{velocity of ultrasound in crystal material}}{\text{wavelength}}$

(ii) wavelength = 2 x thickness (t)

Therefore, frequency (f) = $\dfrac{\text{velocity (v)}}{2t}$

or t = $\dfrac{v}{2f}$

Example

The velocity of ultrasound in a commercial preparation of lead zirconate titanate, a commonly used piezoelectric ceramic material, is 4,000 metres per second. If a vibration frequency of 5 MHz were desired, what would be the crystal thickness?

$$\text{Thickness} \quad t = \frac{v}{2f} = \frac{4{,}000 \text{ m/s}}{2 \times 5 \times 10^{6}/\text{s}}$$

$$= 4 \times 10^{-4} \text{ m}$$

$$= 0.4 \text{ mm.}$$

In practice, crystal thicknesses for diagnostic ultrasound transducers range typically from 0.1 mm for high frequencies to 1.0 mm for low frequencies. **The thinner the crystal, the higher the frequency.**

The size (diameter) and shape of the crystal have an effect on the shape of the ultrasound beam as it travels outwards from the transducer. This is discussed in Chapter 5.

2.3.1.1 Electrical connections

The front and back surfaces of the crystal are coated with thin films of electrically conducting material to facilitate connections to the electrodes which supply the potential difference for pulsing the crystal. The back electrode serves as the live connection, while the front electrode is earthed to protect the patient from electrical shock. In addition to pulsing the crystal during the generation of ultrasound, the electrodes also serve to pick up the piezoelectric signal generated when returning echoes strike the crystal. The front side of the transducer, which makes direct contact with the patient, is covered with an electrical insulator.

2.3.1.2 Backing material

The backing block behind the crystal is made of a material which absorbs ultrasound heavily. Absorption of ultrasound is discussed in Chapter 3. The purpose of the backing block is to absorb ultrasonic energy transmitted back into the transducer, and hence to quickly damp the oscillations of the crystal following a pulsation. This is important in pulsed -wave techniques in which the transducer sends out a short burst of energy (a pulse) followed by a comparatively much longer "listening" phase during which the transducer does not emit ultrasound but is instead tuned to detect returning echoes. Damping also controls the pulse length, which in turn affects image resolution (see Chapter 8). Transducers for continuous wave ultrasound are allowed to vibrate freely at the resonant frequency, and do not require damping.

2.3.1.3 Acoustic insulator

The acoustic insulator prevents vibrations originating from the crystal from being transmitted into the transducer housing. It also insulates the crystal from extraneous sources of ultrasound. It should be made of a material which transmits ultrasound poorly, such as rubber.

2.3.1.4 Transducer housing

The internal components of the transducer are covered by a robust housing.

CHAPTER 3

INTERACTION OF ULTRASOUND WITH MATTER

In order to use ultrasound for either diagnostic or therapeutic purposes, a beam of ultrasound must be directed into the tissues of the subject over a selected area of interest. The ultrasonic energy will then interact with the tissues along its path. In this chapter, the macroscopic processes of interaction between ultrasound and matter are considered. The interaction processes are influenced by the characteristics of the ultrasound wave, as well as the physical properties of the tissues through which the beam passes.

3.1 Acoustic impedance.

Different materials respond differently to interrogation by ultrasound, depending on the extent to which their medium particles will resist change due to mechanical disturbance. This medium property is referred to as the **characteristic acoustic impedance** of a medium. It is a measure of the resistance of the particles of the medium to mechanical vibrations. This resistance increases in proportion to the density of the medium, and the velocity of ultrasound in the medium. Acoustic impedance, Z, may be defined as the product of medium density and ultrasound velocity in the medium.

Eq. 3.1. $\quad Z = \text{density} \times \text{velocity}$

3.2 Acoustic boundaries

Positions within tissue where the values of acoustic impedance change are very important in ultrasound interactions. These positions are called **acoustic boundaries**, or **tissue interfaces.** For example, urine in the bladder will have an acoustic impedance value which differs from that of the bladder wall, hence their common interface constitutes an acoustic boundary. To a very large extent, the unique features of diagnostic ultrasound as an imaging modality are determined by the nature and distribution of the multitude of acoustic boundaries within the tissues of the body.

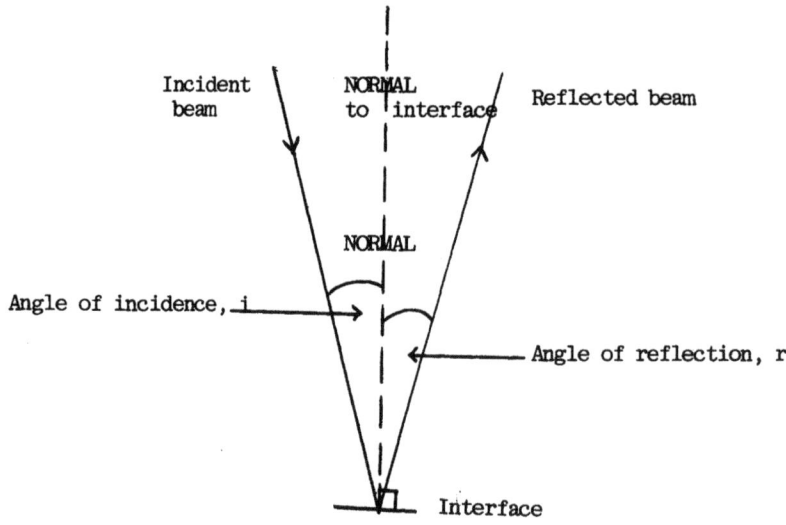

Fig 3.1 For specular reflection, angle of incidence = angle of reflection

In general, the extent to which an acoustic boundary affects a beam of ultrasound incident upon it will depend on the magnitude of the difference between the acoustic impedance values of the two structures on either side of the boundary.

3.3 Reflection of ultrasound

The most important single interaction process for purposes of generating an ultrasound image is reflection. When a beam of ultrasound strikes an acoustic boundary, part of the beam energy is transmitted across the boundary, while some is redirected backwards, or reflected. Two types of reflection can occur, depending on the size of the boundary relative to that of the ultrasound beam, or on irregularities of shape on the surface of the reflector. **Specular reflections** occur when the boundary is smooth and larger than the beam dimensions, while **non-specular reflections** occur when the interface is smaller than the beam.

3.3.1 Specular reflections

A specular reflector is an interface whose diameter is larger than one wavelength of the ultrasound beam. For this type of reflection, a simple law similar to that governing the reflection of light is obeyed. In Fig 3.1 the angle, i, between the incident beam and the perpendicular direction to the interface is called the **angle of incidence**. The line perpendicular to the interface is also called the **normal** to the reflecting surface. The reflected ultrasound will be on the opposite side of the normal in relation to the incident beam, and the angle it makes with the normal is called **the angle of reflection**, r.

For specular reflection: Angle of incidence = angle of reflection.

The reflected beam is referred to as the echo (by analogy with audible sound). The probability that an echo will go back to the transducer and be detected increases as the angles i and r decrease. Detection of an echo by the same transducer producing the incident beam requires that the angle of incidence be small, typically less than about $3°$.

The intensity of an echo due to specular reflection depends on the angle of incidence as well as the difference in acoustic impedance values of the two media forming the boundary. This difference in Z value is also called the **acoustic mismatch**. The most useful specular reflection takes place when the ultrasound beam strikes a reflector at $90°$ to the surface of the boundary. This is referred to as **normal incidence**. The angles i and r are then equal to zero, and the echo goes straight back with a high probability of being picked up by the transducer. In this special case of specular reflection, the echo intensity in relation to the intensity of the ultrasound beam incident upon the boundary is given by the relation:

$$\text{Eq 3.2} \quad \frac{I_r}{I_i} = \frac{(Z_1 - Z_2)^2}{(Z_1 + Z_2)^2}$$

I_r = intensity of reflected echo.
I_i = intensity of incident beam at the boundary.
Z_1 = acoustic impedance of first medium
Z_2 = acoustic impedance of second medium.

The ratio Ir/Ii is called the **reflection coefficient**. It represents the proportion of beam intensity that is reflected from the interface. For a given intensity at the source, echo sizes will vary in proportion to this ratio. It can be inferred from Eq. 3.2 that it is the **difference** in Z-value at an interface which determines the reflection coefficient. A large change of Z value at an interface makes the reflection coefficient large and therefore gives rise to a large echo, whereas small changes of Z produce small echoes. This inference is of the greatest significance in ultrasonic imaging. Table 3.1 shows values of Z for some materials of interest in clinical ultrasound.

TABLE 3.1: VALUES OF ACOUSTIC IMPEDANCE FOR SOME MATERIALS

Material	Acoustic impedance (Rayl)
Muscle	1.70
Fat	1.38
Brain	1.58
Kidney	1.62
Liver	1.65
Blood	1.61
Soft tissue (average)	1.63
Water	1.48
Bone	7.80
Air (NTP)	0.0004

Attention should be focused not so much on the absolute values of Z, or the units in which they have been expressed, but rather on comparisons of the values. In this regard, the following important observations can be made by examining the data in Table 3.1.

(i) The values of Z for the soft tissues are quite similar to one another. We conclude that **reflections at boundaries between soft tissues will give rise to generally small echoes.**

(ii) The Z-value for air (and for other gaseous materials) is much lower than the soft tissue average. We conclude that **reflection from a soft tissue/gas interface gives rise to a very large echo.**

(iii) The Z-value for bone is several times higher than the soft tissue average. Therefore, **a reflection from a soft tissue/bone interface produces a large echo.**

Table 3.2 shows the percentages of ultrasonic energy reflected normally from different boundaries.

TABLE 3.2: PERCENT REFLECTION OF ULTRASONIC ENERGY FOR NORMAL INCIDENCE AT VARIOUS BOUNDARIES

Boundary	**% reflection**
Muscle/fat	1
Kidney/fat	0.6
Bone/muscle	41
Bone/fat	49
Soft tissue/air	99.9
Soft tissue/water	0.2

3.3.1.1 Implications for diagnostic ultrasound

3.3.1.1.1 Negative role of gas

When a very large proportion of the ultrasonic energy incident upon a boundary is reflected, the residual intensity becomes too low for interrogation of structures lying beyond the boundary. **The near total reflection of ultrasound at boundaries between gas and other materials makes gas a barrier to the transmission of ultrasound.**

3.3.1.1.2 Need for coupling gel

It is necessary to apply suitable gel or oil between the transducer surface and the patient's skin in order to exclude air.

3.3.1.1.3 Acoustically homogeneous media

No echoes are seen within compartments in which the Z-value remains constant. The reflection coefficient is zero throughout such compartments and therefore there are no acoustic interfaces within. In clinical ultrasound, **acoustic homogeneity is observed in liquid cavities as clear, echo-free zones.**

3.3.1.1.4 Transducer matching layer

The front face of the transducer, between the piezoelectric crystal and the skin, has a matching layer made of a material of suitable Z and thickness which serves to reduce the acoustic mismatch between the crystal material and soft tissue. The Z-value of the matching layer is chosen to lie between that of the crystal and soft tissue. This reduces the reflection of ultrasound that would be experienced at the crystal/soft tissue interface in the absence of the matching layer, thus improving the transmission of ultrasound into the patient, and of the returning echo into the transducer.

3.3.2 Scattering of ultrasound (non-specular reflections)

When the reflecting interface is irregular in shape, and its dimensions are smaller than the diameter of the ultrasound beam, the incident beam is reflected in many different directions. This is known as **non-specular reflection, or scattering**. The direction of scatter does not obey the simple law of reflection as in specular reflection, but depends instead on the relative sizes of the scattering target and the ultrasound beam diameter. Therefore, energy incident upon the small target at quite large angles of incidence will have a chance of being detected at the transducer. The dimensions of the interface should be about one wavelength of the ultrasound beam or less for scattering to occur. It will be recalled that the wavelengths for typical diagnostic beams are 1 mm or less. Within the organs, there are many structures which have dimensions of less than 1 mm, and so scattered ultrasound provides much useful information about the internal texture of organs. Scattered echoes are much weaker than specularly reflected echoes, but the high sensitivity of modern ultrasound equipment makes it possible to utilize information from scattered ultrasound for imaging. **Scattering shows very strong frequency dependence, increasing rapidly as the frequency of ultrasound is increased.**

3.4 Refraction of ultrasound

Refraction is a change of beam direction at a boundary between two media in which ultrasound travels at different velocities. It is caused by a change of wavelength as the ultrasound crosses from the first medium to the second while the beam frequency remains unchanged. We recall that:

$$\text{Velocity} = \text{frequency} \times \text{wavelength}$$

Therefore, when velocity changes but frequency remains the same, the wavelength must undergo change.

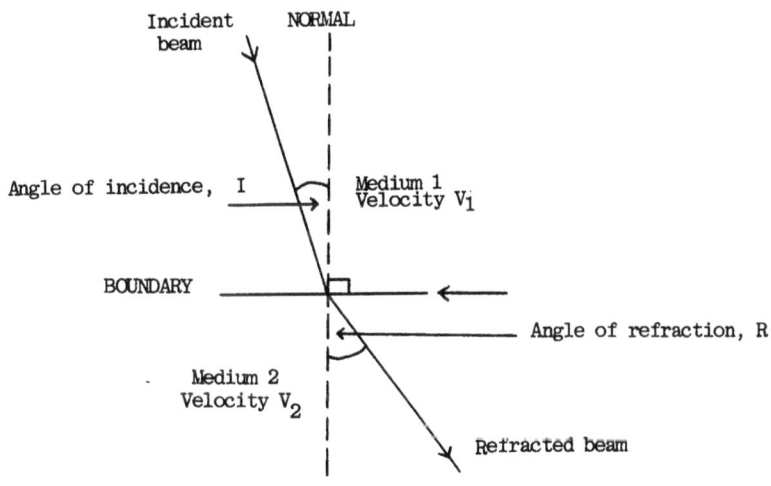

Fig 3.2 Refraction of ultrasound

Fig 3.2 illustrates the phenomenon of refraction. It occurs when the angle of incidence at the boundary is not zero. In the case of normal incidence, part of the beam energy is reflected directly backwards, and the remaining energy is transmitted into the second medium without directional change. At any other angle of incidence, the transmitted beam is deviated from the original direction of the incident beam, either towards or away from the normal, depending on the relative velocities of ultrasound in the two media. The relationship between the angle of incidence and the angle of refraction is governed by Snell's law as in optics:

$$\frac{\sin I}{\sin R} = \frac{V_1}{V_2} \quad \text{(Snell's law)}$$

I = angle of incidence

R = angle of refraction

V_1 = velocity of ultrasound in medium 1

V_2 = velocity of ultrasound in medium 2

Unlike reflection, refraction does not contribute usefully to the process of image formation. However, the deviation of ultrasonic energy into new directions contributes to loss of beam intensity. Refraction may also be the cause of artefacts. (See Chapter 8).

3.5 Absorption of ultrasound

Absorption is the process by which energy in the ultrasound beam is transferred to the propagating medium, where it is transformed into a different form of energy, mostly heat.

The medium is said to absorb energy from the beam. The extent of absorption in a medium is affected by three main variables. These are:

(i) the viscosity of the medium

(ii) the relaxation time of the medium

(iii) the beam frequency.

Viscosity is a measure of the frictional forces between particles of the medium as they move past one another. The greater these frictional forces the more heat generated by the vibrating particles. Therefore, **absorption of ultrasound increases with increasing viscosity.**

Relaxation time is a measure of the time taken by medium particles to revert to their original mean positions within the medium following displacement by an ultrasound pulse. Its value is characteristic of the medium. When the relaxation time is short, vibrating particles are able to revert to their original positions before the next disturbing pulse, but when it is long, the next pulse may encounter the particles en route before they are fully relaxed. The new compression and the particles may then be moving in opposing directions, thus resulting in additional dissipation of energy from the beam. Therefore, **the longer the relaxation time of a medium, the higher the absorption of ultrasound.**

The frequency of vibrations affects the amount of heat generated through both the viscous drag and the relaxation process. Higher frequency means that medium particles move past each other at an increased rate, thus generating more frictional heat. Increased frequency also reduces the probability that, following an ultrasonic pulse, vibrating particles will have reverted to their equilibrium positions before the next disturbance, thereby increasing energy absorption as the new wave moves in opposition to the relaxing particles. We conclude that **absorption of ultrasound increases with increasing beam frequency.**

Rule of thumb: In soft tissues, absorption of ultrasound increases in direct proportion to the beam frequency.

The significance of the process of absorption in diagnostic ultrasound is that energy is removed from the beam, leaving less energy available for examination of tissues lying beyond the absorbing medium. When there is total absorption of beam energy, no diagnostic information can be obtained in the shadow of the absorber. The absorbing object is said to cast an **acoustic shadow** behind it. Among materials of biological interest, **bone absorbs ultrasound much more strongly than the soft tissues.**

3.6 Beam divergence and interference

Divergence of an ultrasound beam describes the spreading out of the beam energy as it moves away from the source, while interference refers to the manner in which different parts of the wave (called wave fronts) interact with each other. The theoretical explanation for the spreading out of beam energy is provided by Huygens' principle of wavelets. The reader is referred to more detailed texts on wave theory for further reading on this topic. Divergence affects the intensity of the beam both axially (along the beam direction) and laterally (perpendicularly to the beam direction). Interference can result in either a strengthening or a weakening of the wave, depending on the phases (positions in the wave cycle) of the interacting wave fronts.

In diagnostic ultrasound the dimensions of the ultrasound beam, and the manner in which it diverges, has a great influence on image resolution, and on the tissue depths that can be investigated using the beam. This important subject is covered in greater detail in Chapter 5.

3.7 Attenuation of ultrasound in tissues

The interaction processes discussed in this chapter have the effect of progressively diminishing the intensity of the ultrasound beam as it passes through the tissues. The tissues are said to **attenuate** the ultrasound beam. The term attenuation should be distinguished from absorption. Whereas absorption relates to the conversion of ultrasonic energy into thermal energy within a medium, attenuation refers to the total propagation losses that result in a reduction of the beam intensity. These losses include those due to reflection, scattering, refraction, and absorption. In other words, absorption is but one of the factors which contribute to attenuation.

3.7.1 Ultrasonic half value thickness (HVT)

The extent of beam attenuation depends on both the beam frequency and the properties of the propagating medium. The attenuation in a specified medium may be quantified in terms of the ultrasonic half value thickness (HVT). **The HVT of a beam of ultrasound in a specified medium is the distance within that medium which reduces the intensity of the beam to one half of its original value.**

Table 3.3 shows values of ultrasonic HVT in different materials for beam frequencies of 2MHz and 5MHz.

TABLE 3.3: ULTRASONIC HALF VALUE THICKNESSES FOR DIFFERENT MATERIALS AT FREQUENCIES OF 2MHz AND 5MHz.

Material	HVT (cm) at frequency of 2 and 5 MHz	
	2 MHz	5 MHz
Muscle	0.75	0.3
Blood	8.5	3.0
Brain	2.0	1.0
Liver	1.5	0.5
Soft tissue (average)	2.1	0.86
Water	340	54
Bone	0.1	0.04
Air (NTP)	0.06	0.01

3.7.2 Acoustic windows and acoustic barriers

Examination of the data in Table 3.3 shows that the HVT values for water (and for other liquids) are large, whereas those for bone and gas are very low. This means that ultrasound can travel long distances in liquids. The reason for this is that clear liquids are acoustically homogeneous - there are no acoustic boundaries within them. This eliminates those attenuation processes which depend on changes at boundaries (including reflection and refraction). Overall attenuation is therefore low. **Compartments which allow ultrasound to pass readily through them are referred to as acoustic windows.** Liquid cavities within the body, containing bile, urine, amniotic fluid, cerebrospinal fluid, and so on, fit this description. On ultrasonic images, they appear as clear, echo-free zones. It should be noted, however, that through the process of absorption, some attenuation of ultrasound still takes place within liquid cavities.

Table 3.3 also reveals that bone and gas impede the flow of ultrasound. Bone absorbs heavily, while gas boundaries reflect almost totally. The presence of fat also impedes the transmission of ultrasonic energy, for a variety of complex reasons. Such materials are referred to as **acoustic barriers.**

3.7.3 Effect of beam frequency

Attenuation of ultrasound increases rapidly with increasing beam frequency. This is because the process of absorption increases approximately in direct proportion to frequency, while the probability for beam scattering increases even more rapidly with increasing frequency. The differences in HVT values at 2 MHz and 5 MHz in Table 3.3 illustrate the effect of frequency on overall attenuation.

In ultrasonic imaging, the rapid increase in attenuation as the beam frequency is increased implies that very high frequencies cannot be employed to examine long distances in tissue. **High frequencies reduce beam penetration**. For reasons of improved image quality (see Chapter 8), use of high frequencies would be desirable, but reduced beam penetration sets the upper limits of frequencies used in diagnostic ultrasound at about 15 MHz.

CHAPTER 4

INTENSITY OF ULTRASOUND

Reference has been made in the preceding chapters to the term "intensity" in relation to ultrasound, without giving it explicit scientific meaning. In this chapter, the concepts of intensity and power of ultrasound are discussed in some detail.

To understand the meaning of intensity, we recall that an oscillating source of ultrasound in contact with tissue transfers its mechanical energy to the particles of the tissue medium, causing them to vibrate. The medium particles then possess energy by virtue of their motion.

Intensity is a measure of this energy. It represents the vigour of mechanical vibrations of the medium particles. Different physical parameters may be used to express this vigour. These include particle displacement, particle velocity, particle acceleration, and particle pressure. Each of these parameters varies in time and in space within the medium, and so does the intensity. Intensity may be expressed either as an absolute measurement, or using a relative scale.

4.1 Absolute measure of intensity

On the absolute scale, intensity is expressed as the rate of flow or energy per unit area.

Definition : The intensity of a beam of ultrasound at a point is the amount of energy passing through unit cross-sectional area perpendicularly to the beam per unit time at that point.

<u>Units</u>: The following units are commonly used to specify absolute intensities in clinical ultrasound.

>Joule (J) for energy
>Seconds (s) for time
>Square centimetre (cm^2) for area.

Using these units, intensity is expressed in joules/second/square centimetre. joule/second represents the rate of flow of energy and is given the special name **watt**.

$$1 \text{ watt (W)} = 1 \text{ J/s}$$

Therefore intensity can be specified as watts/square centimetre (W/cm^2)

The **power** in a beam of ultrasound is the total energy passing over the whole cross-sectional area of the beam per unit time. If the intensity is uniform over the plane of interest, then

$$\text{Power} = \text{intensity (W/cm}^2\text{)} \times \text{area (cm}^2\text{)}$$

When the intensity is not uniform, then the spatial variations within the beam must be taken into account.

The units of power are joules/second, or watts.

Knowledge of the absolute intensity of ultrasound is required for two reasons. First, the output intensity of an ultrasound instrument affects its sensitivity, and hence signal sizes. Secondly, when we wish to assess the potential biological consequences of exposure to ultrasonic energy, we must have knowledge of the amounts of energy actually dissipated in tissue.

It is common for manufacturers to provide information on the output powers of ultrasound instruments. However, the time variations of intensity complicate the power specification. In particular, the temporal differences between pulsed wave and continuous wave applications makes it necessary to quote different quantities such as the peak power and the time averaged power. For example, the peak power for a pulsed wave instrument could be as high as a thousand times the time averaged power. Fortunately, the peak power is applied over very short durations during the total exposure.

4.2 Relative intensity

On the relative scale, the intensity at a point of interest is compared to that at some defined reference point, and expressed in units called **decibels (dB)**.

Definition : The intensity, I, relative to a reference intensity Io, is defined as:
$$\text{Relative intensity (dB)} = 10 \log_{10} (I/I_o)$$

The dB values will be positive if the intensity of interest, I, is larger than the reference intensity, Io, and negative if I is less than Io. The choice of the reference intensity is arbitrary, but must be defined. For example, I could be the intensity at some point of interest in tissue, and Io the intensity on the skin surface.

The logarithmic scale used in the definition has the inherent mathematical characteristic of compressing the intensity scale such that a very wide range of intensities on the absolute scale can be accommodated on a much smaller range on the dB scale.

The dB notation is useful practically because the levels of intensity used in diagnostic imaging are very low and therefore difficult to measure absolutely. The determination of the ratio of one intensity relative to another is easier, since neither of the two has to be measured absolutely.

4.2.1 The 3 dB change

A change of relative intensity by 3 dB is of special significance. For every 3 dB change, there is a change in absolute intensity by a factor of two.

For the reader who is familiar with logarithms, consideration of the following special cases will make this clear.

By definition, intensity in dB $= 10 \log_{10} (I/I_o)$

Special case 1: Intensity at point of interest equals the reference intensity.

When I = Io, decibel level = $10 \log_{10} (1)$

= 10 x 0 dB

= 0 dB

The decibel level at the reference point is equal to zero.

Special case 2: Intensity at point of interest equals half the reference intensity.

When I = 1/2 Io, decibel level = $10 \log_{10} (0.5)$

= 10 x (-0.301) dB

= - 3.01 dB

Reducing the intensity to a half corresponds to a 3 dB reduction in relative intensity.

Special case 3: Intensity at point of interest equals twice the reference intensity.

When I = 2 Io, decibel level = $10 \log_{10} (2)$

= 10 x (+ 3.01) dB

= + 3.01 dB

Doubling the intensity corresponds to a 3 dB increase in relative intensity.

These calculations illustrate that a change in intensity by a factor of 2, be it an increase or a decrease, results in a corresponding change of 3 dB on the relative scale. In Chapter 3, the HVT of a beam of ultrasound was defined. It can now be inferred that **every HVT in a medium reduces the relative intensity by 3 dB.**

Table 4.1 presents data which show this relationship.

TABLE 4.1 RELATION BETWEEN HALF VALUE THICKNESS, DECIBELS, AND ABSOLUTE INTENSITY

No. of HVTs	Relative intensity (dB)	Absolute intensity mW/cm^2
0	-0	1,600 (Reference)
1	-3	800
2	-6	400
3	-9	200
4	-12	100
5	-15	50
6	-18	25
7	-21	12.5
8	-24	6.2
9	-27	3.1
10	-30	1.6

The reference level in these data has been arbitrarily assigned the absolute intensity value of 1,600 mW/cm^2. Note the compressing effect of the logarithmic dB scale, a thousand-fold reduction in absolute intensity from 1,600 mW/cm^2 to 1.6 mW/cm^2 corresponds to a dynamic range of only 30 dB on the relative scale. Any other choice of reference intensity, in place of the 1,600 mW/cm^2, would give the same result.

CHAPTER 5

ULTRASOUND BEAM SHAPE

As a beam of ultrasound travels outwards from the surface of the transducer, the distribution in space of the ultrasonic energy undergoes change. Axially, the intensity of the beam diminishes gradually with distance along the central axis of the beam, while laterally, at any plane perpendicular to the beam direction, the intensity decreases rapidly with distance from the central axis.

Generally, the ultrasound beam spreads out, or undergoes **divergence**, as it moves away from the transducer. The term "**ultrasound beam shape**" is commonly used to describe the manner in which the spatial distribution of the beam changes with distance from the source. The beam shape has very significant effects on the quality of the ultrasonic image, and on the tissue depths that can be usefully interrogated using a particular beam. This section examines the factors which influence ultrasound beam shape, and the associated implications for ultrasonic imaging.

5.1 General shape of the ultrasound beam

It is helpful to consider first the general shape of the ultrasound beam, and to introduce some terminologies used in describing the beam, before examining the various factors which modify this general shape.

The typical manner in which the ultrasound beam spreads out with increasing distance from the transducer, T, is shown in Fig 5.1.

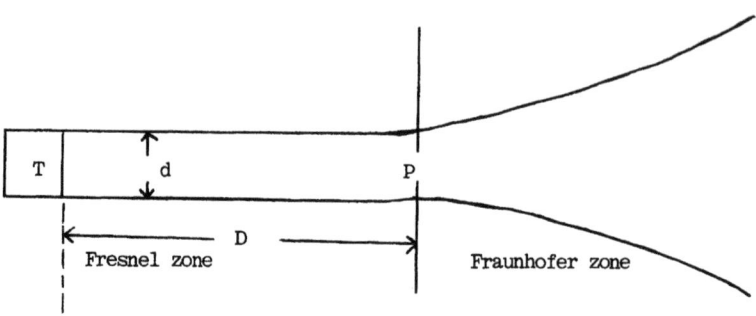

Fig 5.1 General shape of the ultrasound beam

Initially, between T and the plane P along the beam path, the beam is narrow, with a small beam width, d, equal to about the diameter of the piezoelectric crystal. This part of the beam is referred to as the **near field**, or the **Fresnel zone**. Beyond P, the beam spreads out (diverges) over a larger and larger area, with increasing beam widths which result in a rapid deterioration of spatial resolution of the image. This part of the beam is known as the **far field**, or the **Fraunhofer zone**. The distance from the transducer to the plane P is sometimes called the **transition distance** (in reference to the change from Fresnel zone to Fraunhofer zone).

The length, D, of the Fresnel zone, and the beam width, d, at a given plane across the beam, are important parameters which influence, respectively, the practical tissue depth that can be interrogated with the beam, and the spatial resolution in the ultrasonic image (see Chapter 8). The narrow beam associated with the near field is desirable for good spatial resolution. The length of this part of the beam therefore determines the approximate tissue depth which, in practice, can be investigated using the beam.

5.2 Factors influencing beam shape

The shape of the ultrasound beam is affected by:
- the size and shape of the ultrasound source
- the beam frequency,
- beam focusing.

Each of these factors will now be considered separately.

5.2.1 Effect of source size

The size of the ultrasound source affects the beam width, the length of the Fresnel zone, and the angle of divergence beyond the near field. For a transducer in which no focusing is applied, the length, D, of the Fresnel zone is determined by the diameter of the transducer and the wavelength of the ultrasound beam according to the relation:

Eq. 5.1
$$D = \frac{r^2}{\lambda} = \frac{d^2}{4\lambda}$$

where r = radius of the transducer, λ = wavelength of the ultrasound beam
and $d = 2r$ is the diameter of the transducer.

Within the near field, the beam width is approximately equal to the transducer diameter. We infer from Eq. 5.1 that for an unfocused transducer, the length of the Fresnel zone increases rapidly as the beam width (or transducer diameter) is increased.

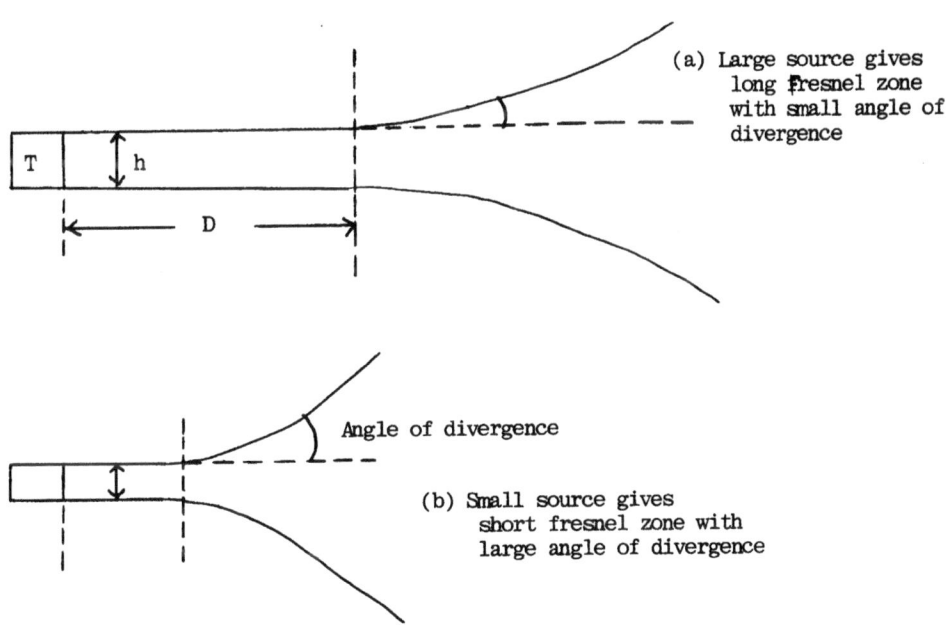

Fig 5.2 Variation of length of fresnel zone and angle of divergence with source diameter

Conversely, the length of the Fresnel zone diminishes rapidly as the transducer diameter is reduced. In addition, a small transducer diameter results in a large angle of divergence beyond the near field (see Figs. 5. 2 (a) and (b)), thereby diminishing the lateral resolution rapidly (see Chapter 8).

An important practical implication of these observations is that, although a narrow beam gives us good image resolution, narrow beams should not be obtained only by making the transducer smaller, as this would also reduce the depth of tissue interrogation. It is for this reason that, in multicrystal transducers where many small crystal elements are used, the crystals are not pulsed individually, but in small groups of neighbouring crystals which then provide an instantaneous beam wide enough to give a sufficiently long length of the Fresnel zone.

In summary, the effects of source size on beam shape are:

(i) a small source provides a narrow beam initially, is associated with a short Fresnel zone, and the beam diverges rapidly beyond the near field.

(ii) a large source provides a broader beam initially, gives a longer Fresnel zone, and the beam diverges more gradually, thus providing better resolution of deeper structures.

5.2.2 Effect of beam frequency

Eq 5.1 can be modified by substituting the wavelength of the ultrasound beam by

$$\lambda = v/f$$

where v = velocity of ultrasound in the transmitting medium,
and f = beam frequency.

Then we get $D = \dfrac{d^2}{4\lambda} = \dfrac{d^2 f}{4v}$ (Eq.5.2)

From this expression, we conclude that the **length of the Fresnel zone increases as the beam frequency is increased.** Also, the angle of divergence beyond the near field diminishes with increasing frequency. The effect of higher frequencies is therefore not only improved image resolution but also an increase in the length of the useful near field. In practice, however, some of this advantage is taken away by increased beam attenuation at higher frequencies (see Chapter 3).

5.2.3 Focusing of the ultrasound beam

The shape of the ultrasound beam can be influenced to varying extents by applying different focusing methods.

5.2.3.1 Shape of the crystal element

The crystal element can be suitably shaped by concave curvature to focus the ultrasound beam (Fig 5.3 (a)). This is an **internal focusing** method, because it is effected in the crystal itself. The degree of focusing will depend on the extent of curvature (radius of curvature) of the crystal.

5.2.3.2 Acoustic lenses

Acoustic lenses made from materials which propagate ultrasound at velocities different from that in soft tissue can be used to focus the beam by refraction. The lens will have concave curvature (Fig 5.3 (b)), and the degree of focusing will be determined by the radius of curvature of the lens.

5.2.3.3 Acoustic mirrors

A concave mirror can be used to focus ultrasound by reflection (Fig 5.3 (c)). Again, the degree of focusing will depend on the radius of curvature.

Acoustic lenses and mirrors provide **external focusing**.

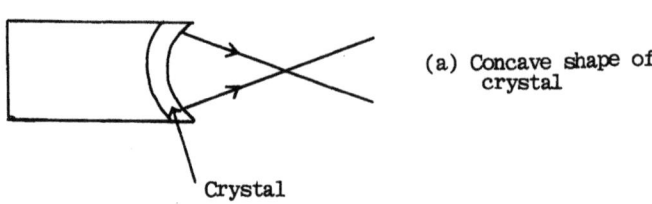

(a) Concave shape of crystal

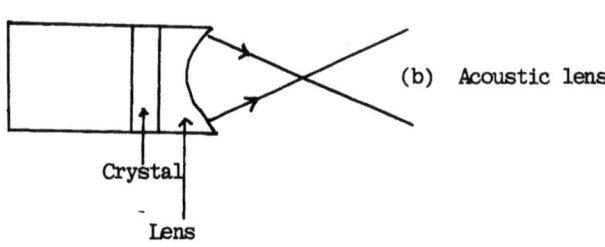

(b) Acoustic lens

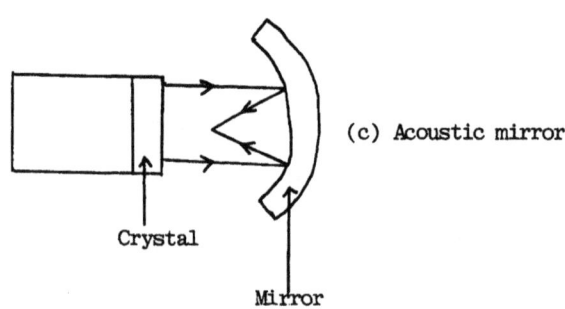

(c) Acoustic mirror

Fig 5.3 Mechanical methods of focusing a beam of ultrasound

5.2.3.4 Electronic focusing

Electronic focusing is employed in multicrystal transducers (see Chapter 7). In such transducers with many crystal elements, movement of the ultrasound beam across the plane of interest in the subject is effected electronically by pulsing small groups of crystal elements at a time. By applying a pulsing programme with carefully controlled time delays between different crystal elements, ultrasound waves from all the crystals in the array can be made to arrive in phase at one particular point (the focus), where they reinforce to produce a high intensity zone. The time delay programme can also be applied during reception of echoes. Electronic focusing offers the advantage of providing **variable focus**, or **dynamic focus**, as opposed to the other methods which provide **fixed focus**. Variable focusing is achieved by altering the time delay programme.

5.2.3.5 Focus of a transducer, focal zone

The **focus**, F, of a transducer is that point along the central axis of the beam which is equidistant in time from all points on the surface of the transducer. The times of flight of the ultrasound waves are equal for all linear paths between the surface of the transducer and F. The waves therefore arrive at F in phase and reinforce each other by constructive interference. Attractive beam properties are associated with the point F: the beam has its narrowest width, greatest intensity, and best spatial resolution.

The focus of a transducer is not sharply defined. Areas within the beam close to F will have properties which will closely match those at F itself. The region around F over which these conditions prevail is called the **focal zone** of the transducer (see Fig 5.4).

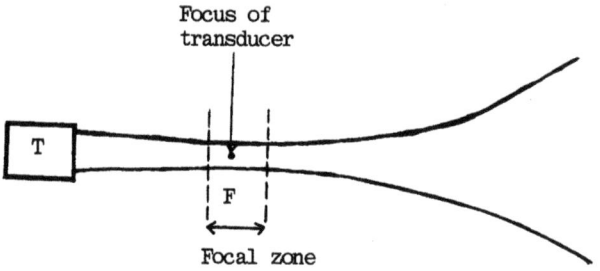

Fig. 5.4 Shape of focused beam, showing the focal zone.

5.2.3.6 Classification of focusing

The degree of focusing may be classified into three categories as follows:

- strong focusing (or short focusing)
- medium focusing
- weak focusing (or long focusing)

In all cases, fixed focusing gives a focal point which is nearer to the transducer than the transition distance (length of the Fresnel zone). Strong focusing brings the focal point very close to the transducer, typically 2 - 4 cm. It achieves a high degree of beam narrowing, but the beam diverges rapidly beyond the focal distance. It can only be applied to transducers for high resolution examinations of small parts. Weak focusing gives a focal point further away from the transducer – typically more than 8 cm – and a gentle divergence of the beam beyond the focus. It is preferred in diagnostic applications because it provides an extended useful, narrow beam.

5.3 Optimization of spatial resolution with tissue depth

The preceding sections of this chapter have shown that the shape of the ultrasound beam is of great significance in ultrasonic imaging. Deliberate efforts are therefore required during transducer design to control the beam shape to suit the desired applications. Generally, a narrow beam would be desirable to maximize spatial resolution of the image, as would be an extended length of the near field in order to facilitate imaging to adequate tissue depths. To achieve these goals requires that the size and shape of the ultrasound source, the beam frequency, and focusing of the transducer, be suitably chosen. In ultrasonic imaging, efforts to enhance one desirable feature quite often works in opposition to another desirable feature. Thus, we have seen that increasing the beam frequency improves image resolution, but also reduces beam penetration due to increased attenuation. A large source of ultrasound at the transducer may extend the useful range of the beam, but it will diminish resolution in the near field. This means that compromises must be made when conflicting interests come into play. The process of balancing opposing interests is referred to as **optimization.** Making the most appropriate choices concerning beam shape characteristics involves optimizing spatial resolution with beam penetration. Special purpose transducers can be designed to suit specific applications. For example, in ultrasonography of small parts, high frequencies can be employed to enhance resolution, because the tissue depths of interest are small, but in examinations of large body sections, lower frequency transducers will be necessary to achieve adequate beam penetration. In the latter case, the demand for high resolution must be compromised to some extent. The optimum choice of frequency would be the highest frequency compatible with the tissue depth requirements. An interesting development in this connection has been the introduction of broad band transducers which offer mixed frequency beams to exploit a bit of the advantages of both low and high frequencies (see Chapter 7).

CHAPTER 6

THE ULTRASOUND IMAGE: GENERATION AND DISPLAY

6.1 Basic principle of the ultrasonic image (see Fig. 6.1)

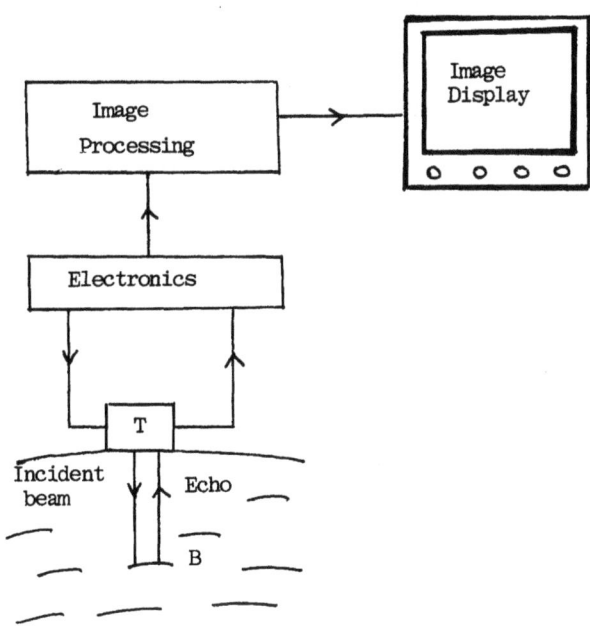

Fig 6.1 Basics of ultrasonic imaging

An ultrasonic transducer, T, sends a beam of ultrasound into the subject over a selected area of interest. At an acoustic boundary such as B within tissue, some of the ultrasound energy is reflected, either specularly or by scattering (see Chapter 3). Under favourable conditions, some of the reflected ultrasound will go back towards T. At the transducer, the returning echo will interact with the piezoelectric crystal and generate an electric signal.

This signal will be electronically processed and measured. The location of its origin at B will be determined.

Ultrasonic imaging involves the mapping of the pattern of echoes reflected from acoustic boundaries within tissues. Characteristic echo patterns are obtained for different tissues. The basic diagnostic parameters to be determined in ultrasonic imaging are:

(i) the size of an echo,
(ii) the distance of echo origin from the transducer.

The ultrasound beam is built by scanning with the beam on the subject. For each position of the ultrasound beam, a set of signals will be recorded along the beam path, corresponding to reflecting boundaries lying at different distances from the transducer. The set of signals produced along one beam path may be referred to as a "**scan line**". It represents single-dimensional information along the beam path. By sweeping the ultrasound beam across the subject ("scanning") in a selected direction, many other scan lines are generated to build a two-dimensional (2-D) image of a plane in the subject. This plane will be defined by the chosen direction of beam sweep. The different methods of sweeping the ultrasound beam are considered in Chapter 7.

6.2 Electronic processing of signals.

The signals generated by the returning echoes at the transducer are electronically processed and organized in computer memory before being displayed on a cathode ray oscilloscope. First, the signals are **amplified** to increase their sizes. The intensity of an echo may be a tiny fraction of the original output intensity of the transducer, hence the piezoelectric voltage it generates at the transducer may be very small. Secondly, echoes returning from different tissue depths must be subjected to **compensation** for attenuation differences. **Time gain compensation (TGC)** is a process of applying differential amplification to signals received from different tissue depths, with echoes originating from longer distances being amplified to a greater extent than those from shorter distances in such a way that similar tissue boundaries give equal sized signals regardless of their depth in tissue. Because the dynamic range of signal sizes may be very wide, the range of signal sizes is **compressed** by using logarithmic amplifiers (the compressing effect of a logarithmic function has been illustrated in the section on decibels in Chapter 4). Finally, signal sizes which are extremely small can be electronically **rejected**. Very small signal sizes have an increased probability of being associated with artefacts. Rejection eliminates all signals whose magnitudes are below a certain threshold level (the rejection level).

The accepted signals are organized in computer memory before being presented to a cathode ray oscilloscope (CRO) for display. Each echo signal is associated with its own intensity level and anatomical position in tissue. Intensity and geometrical coordinates must therefore be assigned to each accepted signal. This information is then read out of memory and displayed as an image on a CRO. Hard copies of the image may then be captured using a suitable recording medium, such as thermal printing paper.

6.3 Echo ranging

One of the essential parameters to be determined in ultrasonic imaging is the distance between the transducer and a reflecting interface. This is done by measuring the time interval between the transmission of a pulse from the transducer and the reception of its echo back at the transducer. To achieve this, two events are electronically "marked", the moment the transducer is pulsed, and the moment it receives the returning echo from the tissue boundary. If d is the distance of a reflecting boundary (B) from the transducer (T), and t is the measured time interval between pulsing and reception, then the ultrasound beam will have travelled a distance equal to 2 d (to-and-fro journey) in the time t (see Fig 6.2). To determine the distance d of the reflector, we use the simple relationship.

Distance = velocity x time

or   2 d = v. t

where v = velocity of ultrasound in the transmitting medium.

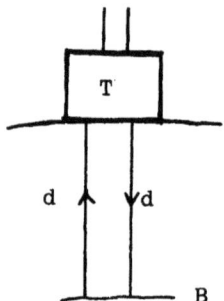

Fig 6.2 Distance travelled by ultrasound beam between pulsing of transducer T and reception of echo is twice the depth d of the reflecting boundary B.

In effect, distance measurements are determined indirectly by measuring electronically the go-return time of ultrasound between the transducer and the reflecting interface. In practice, the electronic time measuring device (a cathode ray oscilloscope) is calibrated to provide distances directly.

Ultrasound systems use the average soft tissue velocity value of 1,540 m/s to calibrate distance measurements. It is interesting to consider the effect of assuming a constant value of the velocity of ultrasound in tissue in the distance calibration of ultrasound equipment, knowing that the velocity does, in fact, vary somewhat in different body tissues. How large are the errors that might be introduced in range measurement by making this assumption? The velocities of ultrasound among the soft tissues do not vary much, and most of them are quite close to their average value. Errors in distance measurement are therefore quite small. For example, the errors introduced in a range measurement of 20 cm during abdominal scanning would be of the order of 2 mm, or 1%, which is small enough to be overlooked. In contrast, the velocities of ultrasound in bone and in gas vary by large amounts from the soft tissue average, and if range measurements were to be made across any substantial distances containing those materials, large errors would be inevitable. Fortunately, this is a matter of academic interest only, since bone and gas are both acoustic barriers (see Chapter 3).

Echo ranging is possible only when ultrasound is used in the pulsed mode, in which the transducer operation is made to alternate between transmission and reception. With continuous wave ultrasound, ranging is not possible because the time interval between pulsing and echo reception cannot be determined.

6.4 Display modes

Once the diagnostic information has been acquired and electronically processed, it has to be displayed for viewing and recording. Different methods are used to display the information acquired from ultrasound examinations. The commonly used modes are outlined in this section.

6.4.1 The amplitude mode (A-mode)

In the amplitude mode, the signals from returning echoes are displayed in the form of spikes on a cathode ray oscilloscope (CRO), traced along a time base (see Fig 6.3). On one axis (vertical axis in Fig 6.3) the amplitude of the signal (magnitude of the voltage pulse) is displayed, and on the other axis (horizontal), the position of the signal on a time scale is represented.

The amplitude of a spike is a relative measure of echo size. Because of the relationship between the distance of a reflector and the time of echo reception, **the position of a spike along the time base is a measure of the distance of the associated reflecting boundary from the transducer.**

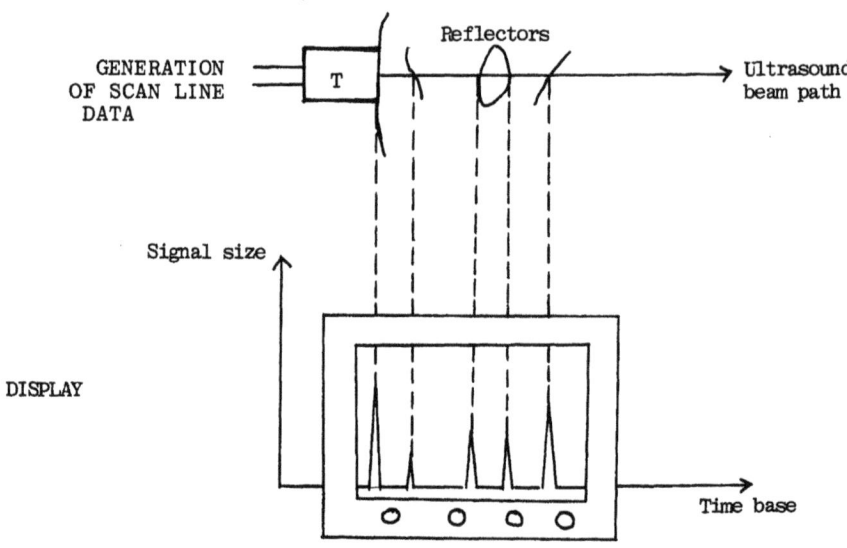

Fig 6.3 Display of information in A-mode: Each spike corresponds to a reflecting interface along the scan line

The A-mode suffers from the limitation of displaying only 1-D information, representing the echoes lying along the beam path. The information does not constitute an image. Additionally, the display has the disadvantage of taking up a lot of space on the CRO in relation to the amount of information that it provides.

6.4.2 The brightness mode (B-mode)

In the brightness mode, signals from returning echoes are displayed as dots of varying intensities. The spike of the A-mode is replaced by a small dot which occupies much less space on the CRO. **The intensity of a dot (the brightness) is a relative measure of echo size,** with large echoes appearing as very bright dots, while at the other extreme non-reflectors appear totally dark. As in A-mode, the signals are presented along a time base on the CRO. **The position of a dot along the time base is a measure of the distance of the associated reflector from the transducer.** For any given position of the beam direction (scan line), a line of dots is displayed on the CRO, corresponding to the 1-D information of reflectors lying along the scan line. When the beam is swept across a selected section of the subject (the process of scanning), different dot lines are created for each scan line. These different dot lines are displayed at different positions on the CRO, displaced laterally from one another, in relation to their corresponding beam positions. **The combined information from different scan lines provides a 2-D image of the cross-section through which the beam sweeps.** One dimension represents depth information, while the other represents lateral variations in the direction of beam sweep. Fig 6.4 illustrates the relationship between the positions of scan lines and the display of dots on the CRO to build the 2-D image.

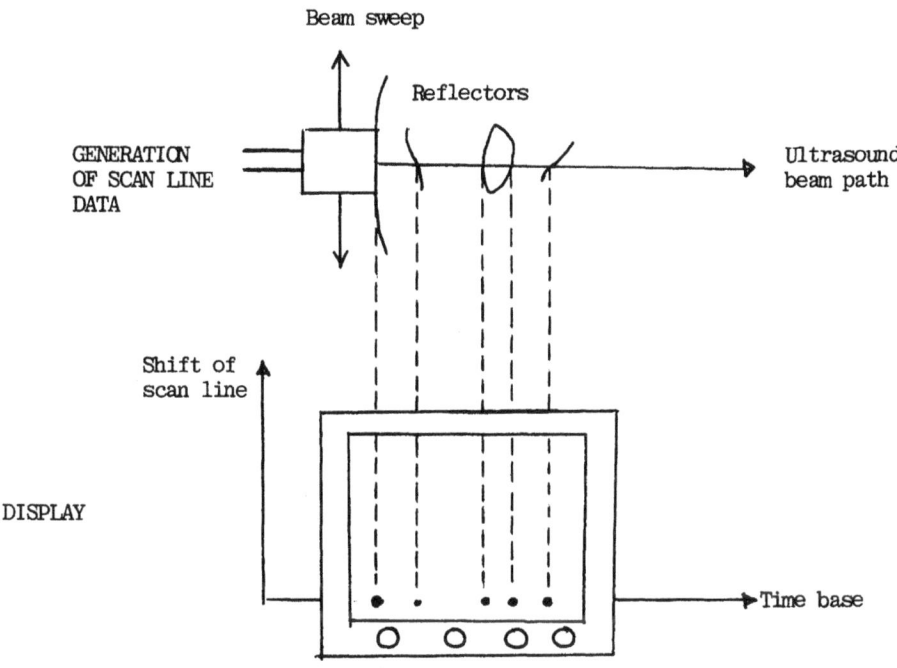

Fig 6.4 Display of information in B-mode: Each dot corresponds to a reflecting interface along the scan line

The speed and the rate at which the ultrasound beam is swept across the subject will determine whether a static or a real-time image will be generated (see Chapter 7).

6.4.3 The motion mode (M-mode)

The motion mode is used to generate an electronic trace of a moving object lying along the path of the ultrasound beam. The transducer is placed in one fixed position in relation to the moving structure. Returning echoes are displayed in the form of dots of varying intensity along a time base as in B-mode. Dots for stationary reflectors will remain in the same positions along the time base, but dots for reflectors which move in the direction of the scan line will change their positions along the time base, because their distances from the transducer will be changing with time. To capture the time variation of moving structures graphically, dot lines obtained at different moments are recorded at different lateral positions on the CRO. This is achieved by applying on the CRO an electronic sweep of the dot lines perpendicularly to the direction of the time base (see Fig 6.5). A time trace of the dot lines along the beam path is thus obtained. It should be noted that the sweep of dot lines on the CRO is achieved by purely electronic means.

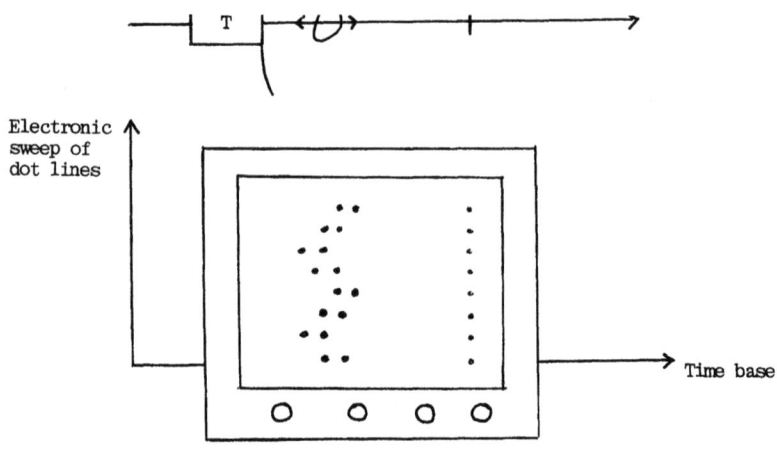

Fig 6.5   M-mode: Traces of dot lines for moving structure M and stationary structure S.

Results: Stationery structures whose dots do not shift positions will trace straight lines perpendicularly to the time base, while moving structures will trace zig-zag or sinusoidal patterns.

The M-mode provides 1-D information along the beam path. It should be noted that for a moving structure to be detected, it must lie along the ultrasound beam path. The M-mode is particularly useful in examining cardiac motion.

6.4.4 Real-time mode

Real-time imaging is rapid B-mode scanning to generate images of a selected cross-section within the subject repetitively at a rate high enough to create the motion picture impression. A rapid succession of images of the same plane are generated and viewed as they are acquired. Although in reality each image in the series represents an independent static image, the effect of rapid acquisition and viewing at rates exceeding about 25 image frames per second creates the impression of continuity in time. This impression arises due to limitations in human perception. We are unable to distinguish in time between events occurring at intervals shorter than about 40 milliseconds - they appear to us to occur "at the same time".

The evolution of ultrasonic imaging into the realm of real-time was a milestone in diagnostic imaging. It is now considered a necessity in the practice of clinical ultrasound. The design of transducers capable of achieving the high framing rates required for real-time imaging is discussed in Chapter 7.

6.4.5 The Doppler mode

Before discussing the Doppler mode as a tool in clinical ultrasound, it is appropriate to introduce the Doppler phenomenon in general. The Doppler effect is observed in the behaviour of sound as well as light. In acoustics, it is associated with relative motion between the source of sound and the receiver, resulting in an apparent difference in frequency between that emitted by the source and that perceived by the receiver. An approaching sound source is perceived to be emitting sound at a higher frequency than it actually is, while a receding source appears to emit at a lower frequency. This situation arises because the wave fronts in the pressure wave of an approaching source are pushed closer together, while the wave fronts in a receding source are pulled further apart (see Fig 6.6).

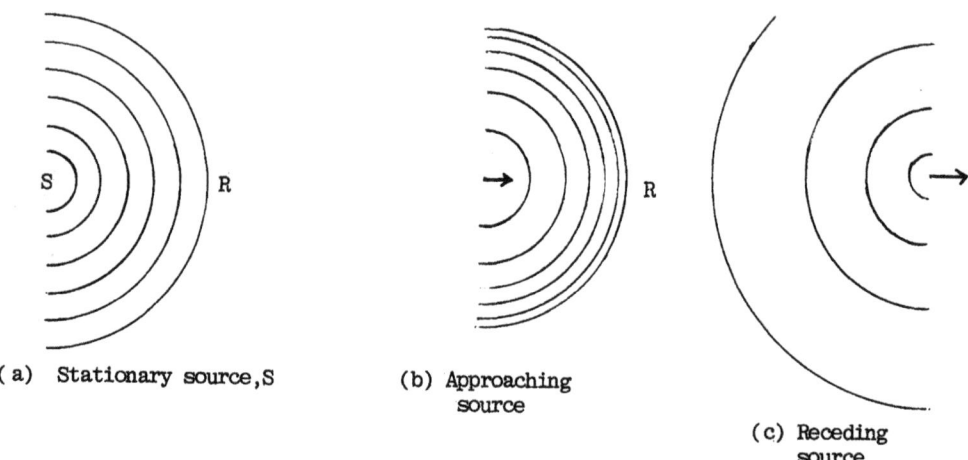

Fig 6.6 Doppler effect: Wavefronts are compressed for approaching source, and decompressed for receding source, as perceived by the receiver, R.

The apparent difference in frequency is called the **Doppler shift**. For a stationary source, the wave fronts are neither compressed nor stretched, hence no shift of frequency is observed. The Doppler shift can be measured and used to:

- detect motion
- determine the direction of motion
- determine the velocity of a moving structure.

In clinical ultrasound, the Doppler mode is used in studies of blood flow and cardiac movements. When a beam of ultrasound emitted by a transducer at constant frequency interacts with a moving acoustic boundary, the boundary will, through the echoes it sends back to the transducer, act as a secondary source of ultrasound for the transducer (effectively, the reflecting boundary becomes the source, while the transducer serves as the observer). Because the boundary is moving, the transducer will detect the echoes with a Doppler shift in frequency, being of higher frequency if the interface is approaching, or of lower frequency if the interface is moving away.

Both continuous wave (CW) and pulsed wave (PW) techniques are used in Doppler ultrasound. In CW Doppler units, the transducer assembly has separate crystal elements for producing the ultrasound beam and for detecting the echoes. One crystal continuously emits and the other continuously receives, it is not possible for one and the same crystal to transmit and detect ultrasound at the same time. By comparing the frequency of the echoes with that of the transmitted beam, it is possible to study motion (see Fig 6.7). The shift of frequency is related to the velocity of the moving reflector, and to the direction of motion. **The greater the Doppler shift, the higher the velocity of the moving structure, and a higher detected frequency implies relative motion towards the transducer, while a lower detected frequency implies a receding reflector.**

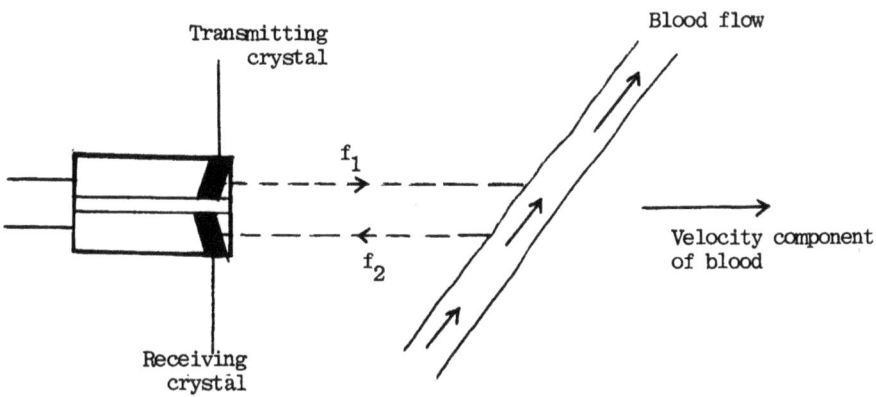

Fig 6.7 Doppler shift in flow studies

The distance of a moving structure from the transducer cannot be determined by CW ultrasound, since the go-return time for the ultrasound beam will not be known. Determination of range requires the use of pulsed beams. More sophisticated Doppler shift equipment utilizes PW ultrasound in conjunction with B-mode scanning to detect movement and determine range, and to produce images of regions of movement.

6.5 Ultrasound equipment and display modes

The general purpose equipment for diagnostic ultrasound will typically be a B-mode scanner capable of generating real-time images. It will be provided with 2 or 3 transducers of different frequency to cater for imaging of various organs. It will have a printer and "image freeze" capability to facilitate printing of the displayed image. The unit may have an M-mode facility incorporated for the study of motion.

Doppler ultrasound is more demanding in terms of both the equipment and the human resources required. It should be offered at specialized centres of patient care where well-trained personnel with specialized skills are available. Where the use of such equipment is appropriate, the ultrasound unit will typically combine real-time imaging with an M-mode option and Doppler facilities of varying complexity.

CHAPTER 7

TRANSDUCERS FOR REAL-TIME IMAGING

The concept of real-time ultrasound was introduced in Chapter 6. Essentially, it involves the generation of images of the same cross-section repetitively and at a rate high enough to create the impression of continuity of events in time. This facilitates the observation of motion in any part of the subject within the cross-section being scanned. The images are generated and erased in rapid succession at rates exceeding about 25 frames per second. The observer simultaneously views the sequence of images, but is unable to resolve them in time, thus the motion picture effect is created. When a permanent record is desired, the image at the particular moment is held still, or "**frozen**", and a hard copy of it recorded (freeze frame). Capturing a record without freezing would cause motional blurring of the captured image.

In order to achieve the high framing rates required in real-time imaging, it is necessary to cause the ultrasound beam to sweep across the section of interest in the subject very rapidly and repetitively. This is the main challenge in the design of equipment for real-time imaging. The transducer and the electronics must be suitably adapted to meet the special demands for high framing rates.

7.1 Transducer design

Transducers for real-time imaging may be classified broadly into two categories: **mechanical transducers and electronic transducers**. In mechanical transducers, the beam sweep is achieved through physical movement of some part of the transducer, usually the crystal element(s), whereas in electronic transducers the beam is swept by electronic activation of crystal elements, without causing the transducer to move physically.

7.1.1 Mechanical transducers

Mechanical transducers are made using either a single piezoelectric crystal or a small group of crystals. A single crystal element may be rocked to perform pendulum motion through a suitable angle. The motion is effected using an electric motor. The angle of swing will define the field of view (see Fig 7.1). The image is triangular in shape and is referred to as a **sector scan**. Each swing of the crystal produces one image frame, and the frame rate is equal to the number of swings per second.

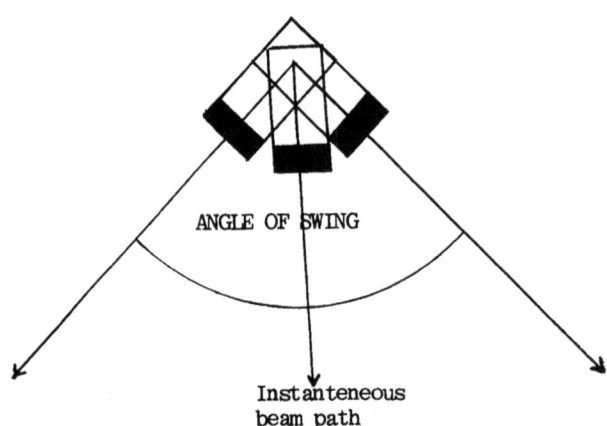

Fig 7.1 Mechanical transducer with a single crystal element performing pendulum motion

Alternatively, the crystal can be driven linearly along the scan section to perform rapid to-and-fro motion. The field of view in this case will be rectangular. Another strategy that has been used with single crystal transducers is to employ an oscillating mirror to swing the beam from a stationary crystal by reflection.

The most common method used in mechanical transducers employs a small group of crystal elements (typically 3 or 4 crystals) mounted symmetrically on a rotating wheel. The wheel is driven by an electric motor to perform circular motion in one direction only. The crystal elements are excited one at a time to provide the ultrasound beam. Each crystal is activated to transmit and receive only as it moves through a predetermined arc which may be referred to as the **active sector**. Outside this arc, the crystals remain inactive. Only one crystal may be active at any given moment. An example of this configuration is illustrated in Figure 7.2.

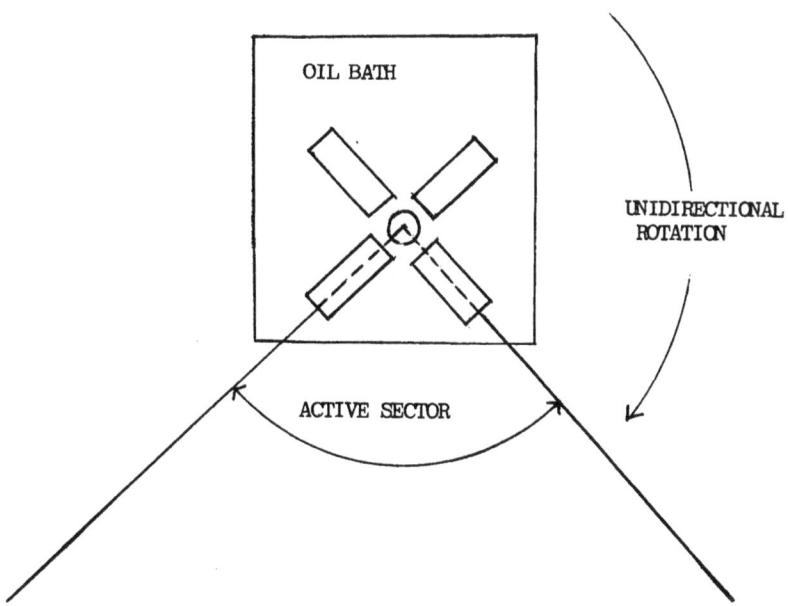

Fig 7.2 Mechanical sector scanner with four rotating crystal elements

For each complete rotation of the wheel, the number of image frames generated will be equal to the number of crystal elements constituting the transducer. The frame rate will further be influenced by the speed of wheel rotation. For example a transducer with 4 crystals performing 10 revolutions per second will produce real-time images at a rate of 40 frames per second.

Since the instantaneous beam is provided by one crystal at a time, the crystal diameters must be appropriately chosen to provide a suitable beam shape (see Chapter 4). Beam shape can further be influenced by focusing. Mechanical transducers employ fixed focusing methods.

At the high speeds used to move the crystal elements, direct contact between the crystals and the patient's skin would be impractical and uncomfortable to both the patient and the operator. To avoid contact scanning, the crystals are housed in a small oil-filled bath. This measure introduces other advantages as well: the oil lubricates the moving parts, the field of view near the skin is improved, and some near field artefacts are transferred from regions of anatomical interest (on the image) to the region of the liquid path (see Chapter 8). Although the ultrasound beam will have to transverse some additional distance across the oil path, beam attenuation is not a major concern because the liquid bath is essentially an acoustic window (see Chapter 3).

7.1.2 Electronic transducers

Electronic transducers are made from a large number of small, identical crystal elements which are acoustically insulated from each other. The crystals are arranged in a suitable geometrical configuration, or an **array,** to provide the desired field of view. Movement of the beam is effected by exciting the crystal elements in an orderly fashion without having to move the transducer physically. The crystal elements may be pulsed individually, one at a time, to provide the instantaneous beam, a pulsing procedure known as **sequential pulsing**. Alternatively, the instantaneous beam may be provided by a small group of crystals excited together. The group is a segment of the array, and such group pulsing is called **segmental pulsing**. The choice between sequential and segmental pulsing is dictated by the need to provide a suitable beam shape that optimizes the conflicting interests of a narrow beam to enhance spatial resolution, on the one hand, and a sufficiently long Fresnel zone that allows for adequate tissue depths to be investigated, on the other hand. In a multicrystal transducer with very many crystal elements, the crystals will, of necessity, be small in size, otherwise the transducer would be too large and bulky. It will be recalled that a small source of ultrasound produces an unsatisfactory beam shape, a short Fresnel zone and rapid divergence in the far field (see Chapter 3). Sequential pulsing of very small crystal elements would therefore give a poor beam pattern. To avoid this problem, segmental pulsing may be employed: a small number of adjacent crystals are excited together as a group to provide the instantaneous beam. Each pulse of ultrasound from this group results in one scan line. The number of crystal elements constituting the group is chosen such that the resulting beam shape will be similar to that from a transducer in which crystals are pulsed individually. (The crystal elements in transducers employing sequential pulsing must be of adequate dimensions to provide a suitable beam shape).

The stepwise shift in the beam sweep across the length of the array is important because it affects the interplay between various image characteristics (see Chapter 8). It is appropriate here to consider this beam shift in relation to the number of scan lines contributing to the image frame, and how this affects the pulsing procedure in the segmental pulsing of multicrystal transducers.

In sequential pulsing, each crystal element generates a scan line on its own. Thus, a transducer with 60 crystal elements will generate an image frame from 60 scan lines. The number of scan lines contributing to the image affects spatial resolution in the direction of the length of the array (lateral resolution). The larger the number of scan lines, the better the lateral resolution. In segmental pulsing, if the groups of crystals were chosen independently and exclusively of each other, the number of scan lines in one sweep of the beam across the array would be much reduced, in relation to the total number of crystal elements in the array. For example, segmenting an array of 100 crystal elements into independent groups of 5 would produce an image frame from 20 scan lines, as compared to 100 scan lines if the same array were pulsed sequentially (we recall, though, that the groups of 5 provide a more practical beam shape). In order to combine the advantages of a suitable beam shape and those of generating the image frame from a large number of scan lines, segmental pulsing is done by sweeping the beam electronically in small steps across the crystal array, by overlapping the sets of crystals instead of grouping them in mutually exclusive segments. We consider the example of a transducer with 100 crystal elements segmented into groups of 5 as the beam is swept from the first to the 100th element. The first scan line is provided by pulsing crystal numbers 1 - 5, the second scan line by crystals 2 - 6, the third by numbers 3 - 7, and so on. The last scan line will come from crystal numbers 96 - 100. In this manner, the shift in the beam sweep is effected in small steps (the lateral shift is equal to the length of one crystal element along the direction of the array), giving rise to a large number of scan lines: a total of 96 in this example. At the same time, the advantages of a large source of ultrasound (the dimensions of 5 crystal elements) are retained.

7.1.2.1 Geometry of multi crystal arrays

Different geometrical configurations have been used in multicrystal arrays. In **linear array transducers,** the crystal elements are arranged in a row. They may be activated either sequentially or segmental. They generate rectangular fields of view (see Fig. 7.3), associated with good visualization of superficial regions.

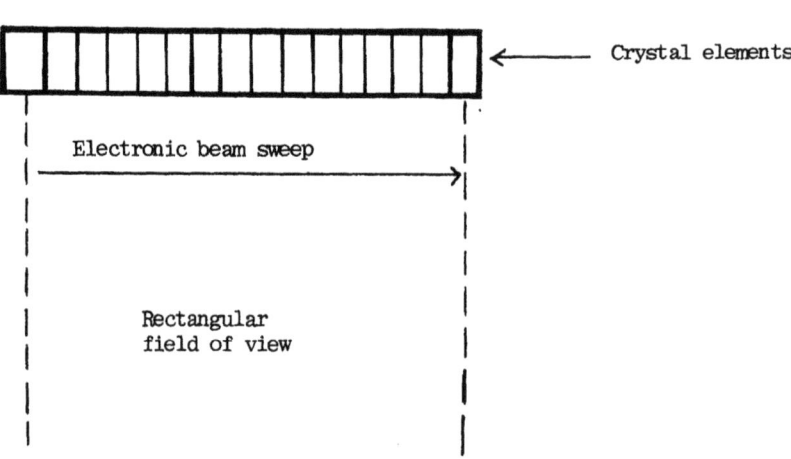

Fig 7.3 Linear array transducer

In an **annular array transducer**, the crystal elements are arranged in concentric rings. For electronic beam sweep, the set of typically 5 - 10 ring elements is pulsed sequentially to move the beam from the innermost ring outwards. The beam can also be moved mechanically by oscillating the whole assembly from side to side, or by reflecting the beam from a moving acoustic mirror.

A **phased array** system may have the same geometrical configuration as either a linear array or an annular array, but the procedure of activating the crystal elements is different. Neither sequential nor segmental pulsing is employed. In phased array transducers, all the crystal elements are pulsed almost instantaneously as one group, excepting for very short time delays between the activations of individual crystal elements. The carefully controlled electronic time delays are programmed into the pulsing of each crystal element to facilitate movement and focusing of the beam (see section on electronic beam focusing in this Chapter). For each pulsing of the array, a phased array system produces only one "scan line" over the whole area of the array. The number of scan lines contributing to the image is increased by sweeping the beam electronically through different directions using the short time delays.

7.1.2.2 Electronic beam focusing

Mechanical systems for focusing the ultrasound beam have been considered in Chapter 5. Transducers for real-time imaging employ some of these mechanical methods, but more significantly they use electronic methods to steer and shape the beam.

Electronic beam focusing relies on the application of very short delays in the pulsing of individual crystal elements within the array to facilitate in-phase arrival of wave fronts from the different elements at the desired focal point. The duration of the time delay for beam focusing is comparatively much shorter than the interval between the pulsing of crystals (or groups of crystals) that is required to gather information and then move the beam to the next line of search across the transducer array. During data collection, time must be allowed for the pulsing of the crystal(s), for the ultrasound beam to travel to the desired tissue depth and back, and for the transducer to detect the returning echo signal. Typically, this sequence of events may take some microseconds (10^{-6}s) to gather information for one scan line. By comparison, the delay times employed in electronic beam focusing are of the order of nanoseconds (10^{-9}s).

Figure 7.4 illustrates a time delay procedure that may be used in the segmental pulsing of a group of 5 crystal elements in order to produce a focal point at F. Outer elements are excited earlier than inner ones in such a manner that wave fronts from all the crystals will arrive in phase at the chosen focal point.

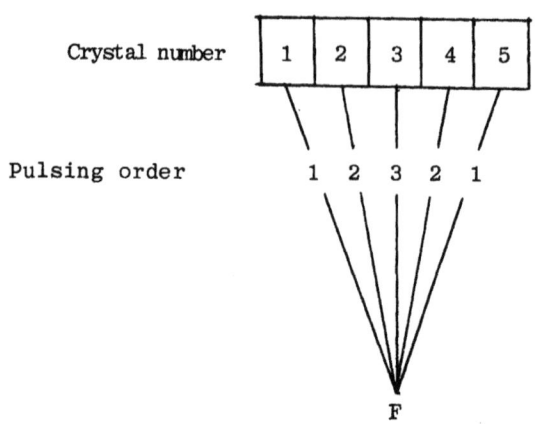

Fig 7.4 Electronic beam focusing. Pulsing sequence is controlled so that wavefronts from each of the crystal elements arrive in phase at the transducer focal point, F.

First, the peripheral crystals 1 and 5 are excited. Ultrasound from these two crystals will travel the longest distance to reach the central axis of the beam produced by the group. After an appropriate time interval, crystals 2 and 4 are excited next. A short delay follows before crystal 3 is fired. The time delays are carefully controlled by electronic circuits to ensure that contributions from all the five crystals in the group will arrive at F in phase to reinforce each other and produce a high intensity zone. The focus can be changed by altering the time delay programme. Variable focal depth is a major advantage of electronic focusing over fixed, mechanical focusing.

7.1.3 Special applications transducers

Some transducers are designed to be used in special applications. This requires that they be specially adapted for the particular tasks. Parameters for special adaptation may include transducer size and/or shape, provision of particular ranges of beam frequency, or some desirable beam focusing capabilities. Examples of such special applications transducers include intra-cavitary probes, cardiac probes, Doppler probes, and high resolution small parts transducers. The reader is referred to specialized articles for further details.

CHAPTER 8

IMAGE CHARACTERISTICS IN CLINICAL ULTRASOUND

The quality of a diagnostic image is of the utmost importance in determining its usefulness. The overall quality of the ultrasound image is the end product of a combination of many factors originating not only from the imaging system but also from the performance of the operator. All the components of the imaging system, including the transducer, the electronics, image processing, display, and recording devices, impact on the ultimate quality of the ultrasound image. It is necessary to emphasize the multifactor nature of image quality in clinical ultrasound because experience shows that the very best of equipment used by an unskilled operator will generate poor quality images, as will unsatisfactory equipment in the hands of a highly qualified operator. In this chapter, the parameters that define the quality of the ultrasound image, and the more important equipment factors which control or affect these parameters, are considered.

A discussion on the quality of the ultrasound image centres around image resolution. **Resolution refers to the ability to distinguish**. Different terminologies are used to express resolution, depending on the parameters to be distinguished. For the ultrasound image, components of resolution include spatial resolution, temporal resolution, and contrast resolution. It is appropriate to consider the factors which affect resolution in some detail, because the optimum choice of these factors very often involves making compromises in the manner of give and take, an essential feature in the optimization of the ultrasonic image.

8.1 Spatial resolution

Spatial resolution describes the ability to distinguish between objects located at different positions in space. In reference to the ultrasound image, spatial resolution is concerned with the ability to distinguish between two reflectors in space. It affects in a major way the capability of the imaging system to depict structural detail. Spatial resolution is divided into two components. **Axial resolution is the ability to distinguish between echoes originating from two reflectors lying one behind the other along the axis of the ultrasound beam** (see Fig 8.1a). It is sometimes referred to as **depth**

resolution. **Lateral resolution is the ability to distinguish between two reflectors situated side by side in a direction perpendicular to that of the ultrasound beam** (Fig 8.1b).

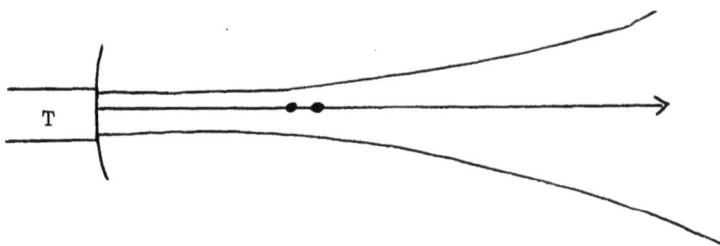

Fig 8.1(a) Axial resolution: Reflectors lie side by side along the beam direction.

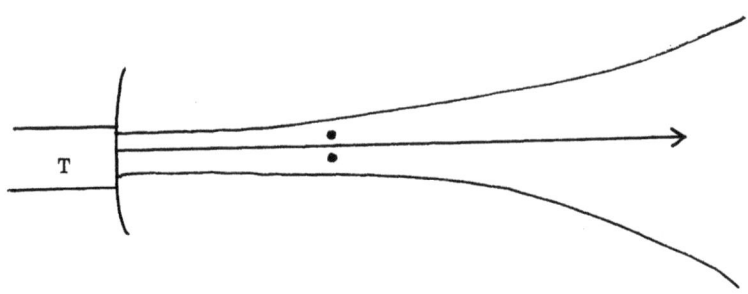

Fig 8.1(b) Lateral resolution: Reflectors lie side by side perpendicular to the beam direction

8.1.1 Axial resolution

Axial resolution is determined by the length of the ultrasound pulse. The variations of wave amplitude with time and distance, and the simple wave parameters of wavelength, frequency, and period, were introduced in Chapter 1. In pulsed wave (PW) ultrasound, the vibrations of the crystals which generate the ultrasound beam are effected in very short pulses, comprising just a few cycles. In between pulses, there are comparatively much longer (non-generating) intervals during which the transducer serves as a receiver. The duration of the pulse is equal to the wave period multiplied by the number of wave cycles in the pulse.

Pulse duration = Period x Number of cycles in pulse.

To achieve the short pulses, the vibrations of the crystal are deliberately damped using materials which strongly absorb ultrasound, otherwise the crystal would continue to vibrate under resonance. The damping material is placed in contact with the back face of the crystal. Figure 8.2 illustrates the variation of wave amplitude with distance for damped oscillations in PW ultrasound.

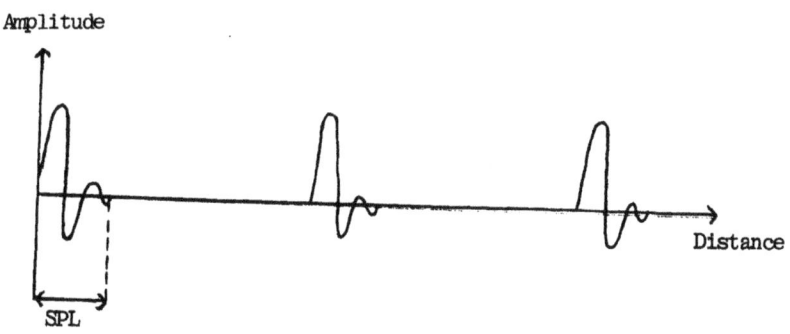

Fig 8.2 Damped oscillations in PW ultrasound.
The amplitude decays with distance

The wavelength remains constant, but the amplitude decays with distance. The oscillations eventually cease altogether after a distance known as the **spatial pulse length (SPL)**. The SPL is the product of wavelength and the number of wave cycles constituting the pulse.

Spatial pulse length = wavelength x number of cycles in pulse

Pulses with identical waveforms are repeated each time the crystal is excited, at a rate known as the **pulse repetition frequency (PRF)**. The PRF represents the number of pulses, or bursts of ultrasonic energy, released by the transducer in one second, and is not the same thing as the vibration frequency of the transducer.

These additional wave parameters are introduced here because they impact in a major way upon image quality. Axial resolution is limited by the SPL. Reflectors closer to one another than half the SPL cannot be resolved.

Axial resolution = 1/2 SPL

The shorter the spatial pulse length, the better the axial resolution. Typically, the SPL of a pulsed beam of ultrasound is of the order of 1 - 3 wavelengths of the beam. Since frequency and wavelength are inversely related, the SPL will decrease with increasing beam frequency. Therefore, **the higher the beam frequency, the better the axial resolution.** Also, because a higher degree of damping quenches the crystal vibrations sooner, the axial resolution will be improved by a greater extent of damping.

The magnitude of the axial resolution can be estimated from knowledge of the beam frequency. For example, a 5 MHz beam travelling through soft tissue (velocity 1,540 m/s) has a wavelength of 0.3mm (see Chapter 1). Assuming that damping restricts the oscillations to 2 wave cycles in each pulse, we have:

$$SPL = 2 \times 0.3 \text{ mm} = 0.6 \text{ mm}$$
$$\text{Hence, axial resolution} = 1/2 \text{ SPL} = 0.3 \text{ mm}.$$

We infer that **when the spatial pulse length is equal to 2 wavelengths of the ultrasound beam, the limit of axial resolution will be equal to the wavelength.**

Spatial resolution can be quantified in two ways. First, as in the example above, we can simply consider the minimum distance between two reflectors that still allows us to distinguish them as being separate entities. We then express resolution in units of distance such as millimetres. Alternatively, spatial resolution can be expressed as the number of line pairs that can be accommodated within the space of 1 mm whilst still recognizing the lines as being separate. The spatial resolution is then expressed in line pairs per millimetre. A large number of line pairs per millimetre corresponds to better resolution (counting the number of line pairs rather than the lines themselves emphasizes the distance between two objects in the concept of spatial resolution, the actual number of lines is one more than the number of line pairs).

The axial resolution does not vary much with tissue depth, as does lateral resolution. Nevertheless, small variations do occur as a result of changes in the frequency spectrum of the beam. The frequency spectrum in the ultrasound beam comprises the nominal or centre frequency with a variable spread of frequencies on either side of the centre frequency (see section on broad-band transducers in this chapter). Because higher frequencies of ultrasound are attenuated more strongly than lower frequencies, at longer distances along the beam path, the beam will comprise higher proportions of lower frequency components. The increased average wavelengths imply longer spatial pulse lengths, hence reduced axial resolution.

8.1.2 Lateral resolution

In ultrasonic imaging, axial resolution is better than lateral resolution, besides showing less variation. This means that lateral resolution is the more limiting aspect of spatial resolution. It is therefore important that the factors affecting lateral resolution be well understood. These factors include:

- beam width
- beam frequency
- scan line density

8.1.2.1 Effect of beam width

Whereas axial resolution is limited by the **length** of the ultrasound pulse, the lateral resolution is limited by the **width** of the pulse. Although it is more common to refer to beam width rather than pulse width, the two are essentially the same, since the ultrasound beam comprises a series of identical pulses released in rapid succession.

Lateral resolution is limited by the beam width in the plane of the reflectors being resolved. Reflectors closer to one another than the beam width cannot be resolved (see

Fig 8. 3a & b). Therefore, **the narrower the beam width, the better the lateral resolution**.

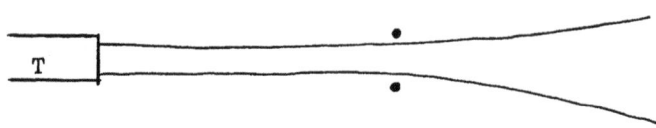

Fig 8.3(a) Small beam width, Reflectors farther apart than the beam width are spatially resolved.

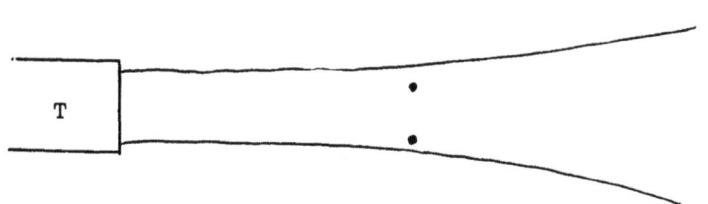

Fig 8.3(b) Larger beam width. Reflectors closer to one another than the beam width are not resolved.

The lateral resolution is highly dependent on tissue depth, since the beam width may vary widely with distance along the beam path (see Chapter 5). It will be evident that beam focusing improves lateral resolution in the near field, and affects its variation along the beam direction. The lateral resolution will be at its best in the focal plane of the transducer.

8.1.2.2 Effect of beam frequency

Frequency affects the beam shape, and hence has a major influence on lateral resolution. The ultrasound beam can be made narrower at higher frequencies. Therefore, **the higher the frequency, the better the lateral resolution**. Also, increased beam frequencies extend the Fresnel zone, although this advantage is partly counteracted by increased beam attenuation at higher frequencies.

8.1.2.3 Effect of scan line density

The methods of acquiring the pulse-echo data from which the ultrasound image is built was discussed in Chapter 6. The image is formed by combining echo information from a large number of scan lines generated by scanning the beam across the selected plane of interest in the subject. The number of scan lines contributing to the image affects lateral resolution. Sampling the tissues at closer intervals improves resolution. Therefore, **the higher the scan line density the better the lateral resolution**. The line density may vary with tissue depth, depending on the type of scanner. In particular, the line density decreases with distance from the transducer in sector scanners, hence the lateral resolution will also diminish with increasing tissue depth (see Fig 8.4). This effect is not present in systems in which the scan lines remain parallel to one another.

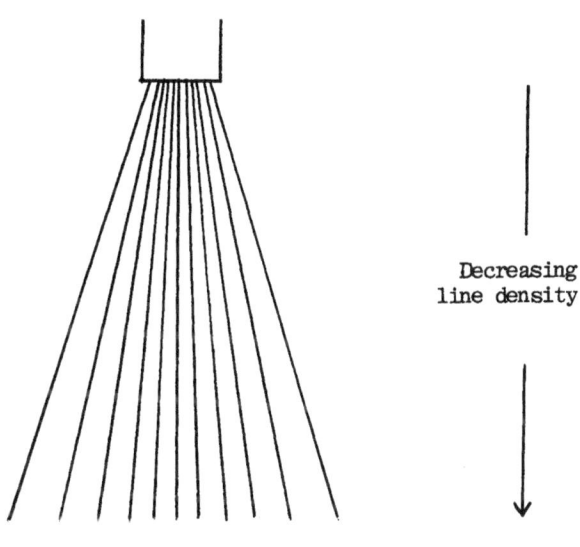

Fig 8.4 Variation of scan line density with tissue depth in sector scanner. Decreasing line density reduces lateral resolution for deeper structures.

8.2 Contrast resolution

Contrast resolution is the ability to distinguish between signal sizes. In the ultrasound image, this translates to differentiating between the intensities of the dots representing echoes of different size at the display. The ability to detect small changes in the characteristic echo pattern of an organ may well depend on the level of contrast resolution. The electronics in the imaging system, and the inherent contrast properties of the display and recording devices, affect the contrast resolution to a large extent. The use of digital image processing techniques is contributing to improved contrast resolution of the ultrasound image.

8.3 Temporal resolution

Temporal resolution is the ability to separate events in time. It is important in real-time imaging. The rate at which image frames are generated and viewed affects the visualization of moving structures. The limit of temporal resolution for the human eye is about 40 milliseconds. This means that events separated in time by more than 40 ms can be visually recognized as having occurred at different moments in time, whereas events occurring within a time interval of less than 40 ms are viewed as taking place "simultaneously". This level of temporal resolution dictates that the framing rate required for real-time imaging to observe moving structures should be 25 frames per second (f.p.s) or more (if one image frame is generated every 40 ms, then in a total time of 1 sec, or 1000 ms, 25 image frames will be generated). Framing rates below about 20 f.p.s are associated with a phenomenon called **image flicker**, which arises from the ability of the eye to distinguish the resulting image frames as being separate in time.

8.4 Limitations due to the velocity of ultrasound

Although ideally it would be good to be able to select each imaging parameter independently of the other parameters in such a manner as to maximize the contribution of each parameter to the enhancement of image quality, this is not possible under all circumstances. Limitations arise due to the finite velocity of ultrasound. In preceding chapters, the velocity of ultrasound in soft tissues was hailed as being fortunately high enough to facilitate echo data collection within very short time periods, therefore making ultrasonic imaging feasible. We must now discuss the constraints that may be imposed upon image characteristics by virtue of the velocity of ultrasound not being higher!

In order to generate one image frame, the ultrasound beam must be made to travel to-and-fro along each of the scan lines contributing to the image, and to the desired tissue depth. The total distance travelled by the beam will increase as the tissue depth is increased, and as the number of scan lines increases. Furthermore, the time available for generating each image frame is restricted by the desired frame rate - the higher the frame rate, the shorter the time must be for each image frame. We introduce the inherent constraints by way of an example.

Hypothetical situation

It is desired to generate real-time images of a body section up to 20 cm tissue depth. The body section contains some rapidly moving structures. A high scan line density of 100 scan lines per image frame is chosen for good lateral resolution, and a high framing rate of 100 f.p.s chosen for good visualization of the moving structures.

Problem

We consider the total distance that the ultrasound beam would need to travel in order to generate the 100 image frames, in a time of 1 second.

Distance per scan line	= 2 x 20 cm = 40cm
Distance per 100 scan lines	= 100 x 40 cm = 4,000 cm
Distance per image frame	= 4,000 cm
Distance per 100 image frames	= 100 x 4,000 cm = 400,000 cm
	= 4,000 m

This distance would have to be covered in a time of 1 second, because the desired frame rate is 100 f.p.s. However, the feat is impossible, because the velocity of ultrasound in soft tissue is only 1,540 m/s!

This example illustrates that the velocity of ultrasound can become a constraining factor in the free choice of imaging parameters that influence image quality. Optimization of image quality often involves making compromises between the various image characteristics to achieve a balance between conflicting interests.

8.5 Optimization of image characteristics

8.5.1 Spatial resolution versus tissue depth

We have seen that higher beam frequencies are associated with better spatial resolution (both lateral and axial). But higher frequencies also suffer more attenuation, thus limiting their useful practical range in tissue. Investigation of small parts to tissue depths of up to about 4 cm can be undertaken using high frequencies of 5 MHz and above in order to maximize spatial resolution. However, it is necessary to use lower frequency beams to achieve increased tissue depths, a measure which compromises resolution. There is a conflict of interest between tissue depth and spatial resolution which necessitates a compromise.

8.5.1.1 Use of multi-frequency transducers

The dilemma of balancing spatial resolution with the requirements of tissue depth is partly addressed by the use of multifrequency transducers. There are two designs of multifrequency transducers.

Broad band transducers

Broad band transducers are designed to provide a wide range of frequencies in the same beam. In a typical ultrasonic transducer, the frequency spectrum comprises the centre frequency, as determined by the crystal thickness, and a narrow spread of frequencies around this centre frequency. This spread may be caused by a number of different factors, including imperfections of crystalline structure, irregularities in the crystal faces, non-parallel crystal faces, wave phenomena, and others. It is possible for the manufacturer to deliberately manipulate some of these factors in order to influence the magnitude of the frequency spread.

A typical frequency spectrum is illustrated in Fig 8.5, which shows a plot of relative beam intensity against frequency. The maximum intensity is observed at the centre frequency f_o. The beam contains frequencies above and below f_o, with variable relative intensities.

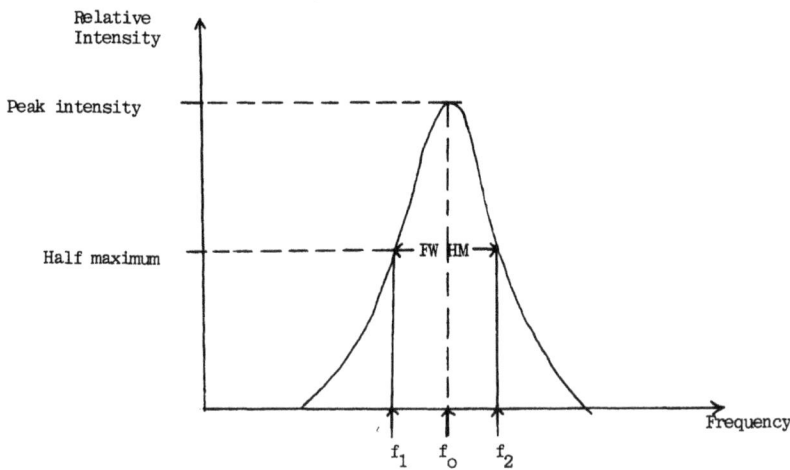

Fig 8.5 Spread of frequencies around the centre frequency, f_o, of a broad-band transducer

The magnitude of the spread of frequencies around the centre frequency f_o is expressed in terms of the **bandwidth**. This may be defined as the range of frequencies present in the beam when the intensity is equal to one half the peak intensity. In Fig 8.5, the bandwidth is represented by the difference in frequency between f_1 and f_2. This difference represents the **full width** of the frequency distribution at half the maximum intensity, and is appropriately called the **full width at half maximum (FWHM)**.

$$\text{Bandwidth} = f_2 - f_1 = \text{FWHM}$$

Broad band transducers, as the name suggests, are designed to offer a large bandwidth. This translates into a wide range of frequencies in the beam, thus combining the advantages of high spatial resolution from the higher frequency components with those of deeper penetration from the lower frequency components.

Multiple centre-frequency transducers

Transducers which offer selectable centre frequency rely on a change of crystal thickness (see Chapter 2). The transducer is designed using a number of crystal layers stratified adjacent to one another. A high frequency beam may be generated by connecting a single layer of the crystal material to the pulsing electrodes, while lower frequencies may be obtained by pulsing the transducer across a sandwich of 2 or 3 adjoining layers of the crystal. Switching from one frequency to another then amounts to changing the effective thickness of the crystal. The lower frequencies will be used for large tissue depths, while the higher frequencies will be for small parts. These transducers are designed to provide narrow bandwidths to maximize the "purity" of the centre frequency (high intensities are then available around the chosen centre frequency).

8.5.2 Frame rates, scan line density, and tissue depth.

Lateral resolution is improved by having a large number of scan lines contribute to the image. Also, high framing rates are desirable in the imaging of fast moving structures as in cardiology. The total number of scan lines generated per second will be limited by the time required to produce one such scan line, depending on the desired tissue depth. Therefore, high framing rates dictate a reduction in the number of scan lines contributing to each image frame, or a reduction in tissue depth. The optimum choice between frame rate, line density, and tissue depth is a matter of compromise. **For a given tissue depth, the higher the frame rate, the lower will be the line density**. A combination of high frame rates with high line density is only feasible for small tissue depths.

8.5.3 Summary of compromises

8.5.3.1 For high lateral resolution:
- beam frequency should be high
- number of scan lines should be large
- tissue depth will be restricted
- frames per second may be limited

8.5.3.2 For large tissue depths:
- low beam frequency is used
- spatial resolution is reduced
- frames per second may be limited.

8.5.3.3 For very high framing rates (rapid motion)

- number of scan lines per image frame may be reduced
- lateral resolution may be reduced
- tissue depth may be restricted.

8.6 Contrast properties of the image display and recording systems.

The oscilloscope which displays the image after it has been electronically processed, and the printer which captures hard copies of the displayed images, are both crucial in determining the quality of the image presented to the viewer. Each has its own inherent contrast, sensitivity, and latitude characteristics which affect both spatial and contrast resolution of the image. In addition, the contrast and brightness controls on the CRO afford further electronic means of manipulating detail perception. Due attention should always be given to the quality of the image display and recording systems used with ultrasound equipment.

8.7 Sensitivity of the imaging system

Sensitivity represents the capability of the imaging system to detect very small signal sizes in relation to the energy expended in the effort to generate the signals. The smaller the minimum detectable signal for a given energy input, the higher is the sensitivity. In ultrasonic imaging, sensitivity determines the ability of the imaging system to detect small echoes, and affects the amount of energy absorbed within the tissues of the subject during an examination. Since a large proportion of diagnostic information is carried by weak echoes, system sensitivity is a critical aspect of ultrasonic imaging.

Many different factors influence the sensitivity of the imaging system, including transducer factors, beam characteristics, signal processing, and image display and recording. The transducer and the electronics will determine the size of the ultrasonic pulse transmitted into the body. The transducer converts electrical energy into mechanical energy at transmission, and mechanical energy into electrical energy at reception of echoes. The efficiencies of these conversions are the most important transducer factors affecting sensitivity. They are determined by a combination of mechanical and electrical properties of the transducer, such as acoustic impedance, impedance matching, damping characteristics, and coupling of the transducer to the pulsing and signal processing electronics. Beam factors affecting sensitivity include the transmitted output power (or reference intensity) of the equipment, the frequency spectrum of the beam, the applied focusing, and attenuation by the tissues. Signal processing involves, among other operations, amplification of signals, compression of signal sizes, and suppression of the weakest of signals. All these operations affect sensitivity. The responses of the display and recording systems to incoming signals depend on the inherent sensitivity characteristics of the display and recording media, such as the fluorescence efficiency of a CRO screen phosphor, or the speed of a photographic emulsion.

8.8 Artefacts in clinical ultrasound

As in other areas of imaging, ultrasound systems can generate misleading information which may misdirect the practitioner towards false interpretation. Artefacts are more likely to occur with older, or simpler models of ultrasound equipment. It is important that practitioners be sensitized to the real possibilities of encountering artefacts in the images they generate and interpret. Awareness is perhaps the most important safeguard against being misdirected.

The major causes of artefacts include multiple reflections across acoustic boundaries, acoustic shadowing due to strong reflectors or absorbers of ultrasound, and poor physical condition of the transducer. Other artefacts may be caused by refraction of ultrasound, wave interference phenomena, or less than perfect mechanical and electrical isolation of crystal elements.

Multiple reflections across tissue boundaries, or between the transducer and tissue boundaries, are known as **reverberations**. They give rise eventually to a signal which corresponds spatially to the total distance covered by the beam, a position at which there may be no actual reflector. Echoes due to this phenomenon are sometimes referred to as "**ghost signals**", because the apparent reflectors do not actually exist. The resulting echo sizes tend to be small due to attenuation along the extended distance traversed by the beam, so in some instances they can be eliminated through electronic suppression. However, it may be difficult to distinguish between ghost signals and genuine echoes from weak reflectors.

Acoustic shadowing occurs beyond heavy reflectors such as gas, or heavy absorbers of ultrasound such as bone, leaving inadequate beam energy for examination of structures lying beyond such structures. This can sometimes be diagnostically useful, as in the case of the classical appearance of shadowing by gallstones (see Fig 8.6), but at other times, it can also lead to difficulties of interpretation.

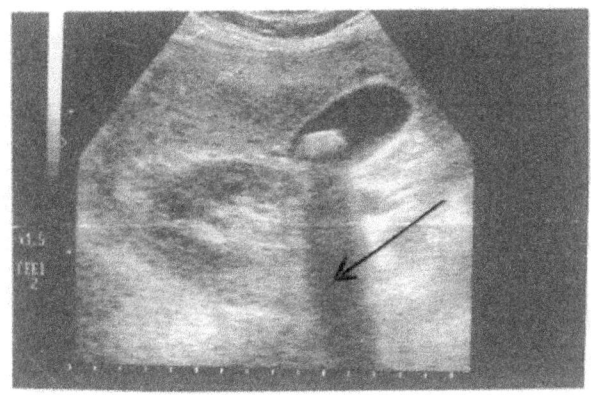

Fig 8.6   Acoustic shadowing by gallstone (arrowed)
(Image courtesy of Dr. Milcah Wambugu,
Department of Diagnostic Radiology,
University of Nairobi)

In a damaged transducer, some crystal elements may have cracks, the vibrations may be poor and irregular, or the damping mechanism may be faulty. These and other transducer faults may generate strange, difficult to account for signals in the images.

Refraction artefacts arise from bending of the ultrasound beam at acoustic boundaries where the velocity of ultrasound changes. When the angle of refraction is large, an apparent displacement of the actual position of a reflector lying beyond the boundary is observed (see Fig   8.7)

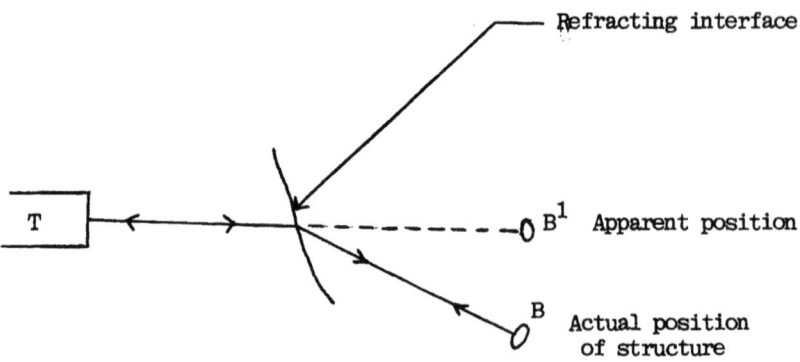

Fig 8.7   Refraction artefacts: The position of a structure B appears to be at the wrong position B^1.

Complex wave interference phenomena in the ultrasonic near field may produce what are known as **side lobes** outside the main beam. These side lobes have intensities which are lower than the useful, main beam. Reflections caused by these side lobes can produce artefacts.

The field of view in the near field may be very restricted in sector scanners, making it difficult to accurately visualize superficial structures. The field of view is improved near the surface when there is a liquid path between the transducer and the surface. Other near field artefacts may also be transferred from regions of interest in the image to the space occupied by the liquid path. This enhances image quality near the surface.

8.9 Quality control

The generation of satisfactory ultrasound images depends on the skills of the operator as well as the performance of the equipment. The former is addressed through personnel training, and the latter through proper selection, care and maintenance of the equipment to ensure optimum performance. Ultrasound equipment contains some delicate components which can easily be damaged physically or electrically. These include the transducers and the computer. Conscious effort must be made to protect the equipment during routine usage. For example, dropping a transducer, or knocking it against other objects, may lead to damage. The accumulation of dust in the electrical components, especially the computer, causes danger to the equipment. Regular servicing of the equipment by qualified technical personnel should be catered for in the operating plans, and implemented. For ultrasound equipment, routine servicing is required about four times a year. Where resources are available, quality control measures can include performance testing of the equipment. Various test objects have been developed for testing ultrasound equipment. However, their use in practice has not gained the same level of acceptance as the quality assurance kits used in other areas of medical imaging. Particularly useful would be phantoms which can be used to assess image resolution, beam penetration, the accuracy of measurement callipers, system sensitivity, and dynamic range.

CHAPTER 9

SAFETY OF DIAGNOSTIC ULTRASOUND

9.1 Energy deposition in tissue

The potential biological effects of exposure to ultrasound arise from the transfer of energy from the ultrasound beam to the biological material. The transfer of energy initiates microscopic particle oscillations within the medium. These oscillations vary in time and in space. They may be characterized by different parameters, including particle displacement, particle velocity, particle acceleration, and particle pressure. Each of these parameters is related to the vigour of particle vibrations, which is a measure of the intensity of the ultrasound beam.

The potential biological consequences of the irradiation of tissues with ultrasound will depend not only on the total amount of energy deposited per unit mass of tissue, but also on the spatial and temporal distribution of this energy. This makes it difficult to uniquely relate the magnitude of biological effects to intensity parameters. Some of the particle oscillation parameters mentioned here may be more significant than others in initiating particular biological consequences. Peak values and time average values may have different impacts on the possibilities for biological effects. In addition, wave parameters such as beam frequency, pulse repetition frequency, and the shape of the pulse, may influence the conditions under which particular effects may be induced. In this regard, there will be major differences between exposure to pulsed wave (PW) and continuous wave (CW) ultrasound. With CW ultrasound, the tissues are irradiated throughout the duration of the exposure, and the total energy deposited in tissue will be much more in comparison to PW irradiation over the same total exposure duration, all other factors being equal. This difference is best expressed by specifying the **duty factor** of a transducer, which is the ratio of the "ON" time to the "ON plus OFF" time, in other words the pulse duration divided by the pulse repetition time. For pulsed wave diagnostic applications, the duty factor is of the order of 0.1%, compared to 100% for CW applications.

9.2 Intensity parameters

It will be evident from the above considerations that the specification of absolute intensity, which is necessary for assessing the biological effects of exposure to ultrasound, is a complex matter. A point-by-point specification of intensity within the irradiated volume of tissue would be such a complex exercise that it is not usually attempted in practice. Instead, the intensity is usually stated in terms of the acoustic beam incident upon tissue. The specification takes different forms, each describing different spatial and temporal conditions. International consensus is desirable on which intensity parameters need to be specified in relation to the assessment of biological hazards, and also on the practical aspects of how these parameters should be measured. Such agreement is necessary to facilitate comparisons in research findings between different centres, and to promote uniformity in the adoption of safety standards. Intensity parameters in common usage include the following:

> Spatial average, temporal average intensity (SATA)
> Spatial average, temporal peak intensity (SATP)
> Spatial peak, temporal peak intensity (SPTP)
> Spatial peak, temporal average intensity (SPTA)

Besides the output intensity of ultrasound equipment, and the mode of operation (CW or PW), the total amount of energy deposited in tissue will vary with the duration of exposure.

9.3 Exposure conditions

For ethical reasons, experimentation with ultrasound for purposes of investigating possible biological consequences are restricted to exposure of animals, of inanimate matter such as water, or of biological samples **in vitro**. It is important to appreciate the differences that usually exist between such experimental irradiations and the diagnostic exposure of human subjects. Firstly, the intensities of ultrasound and the durations of exposure used in these experiments are usually much more than those employed in diagnostic ultrasound. Secondly, the irradiation of small animals such as mice using transducers designed for clinical applications tends to irradiate much larger proportions of volume in the animals in comparison to the human species (for example, a typical diagnostic beam may envelop the whole of the mouse foetus, whereas it would cover only a small part of the human foetus at any one time). The magnitude of potential biological effects can reasonably be expected to vary with the volume of tissue irradiated. Finally, the extrapolation of animal experimental findings to the human case, or of **in vitro** findings to the **in vivo** situation, may present some difficulties and uncertainties. The vast differences in exposure conditions between experimental irradiations and diagnostic applications affects the assessment of potential risks. It has been difficult to generate reliable epidemiological data directly from human subjects undergoing diagnostic examinations. Because of the low levels of exposure, very large sample sizes of subjects would be required to make such studies statistically reliable. Also, as a result of the increased usage of ultrasound worldwide, it is becoming increasingly difficult to obtain control populations not exposed to ultrasound for comparison.

9.4 Consequences of high intensity, prolonged exposures

There are two major consequences of exposure to ultrasound at very high intensities, for long durations, in comparison to the corresponding conditions for diagnostic exposures. These include the elevation of tissue temperature, and a phenomenon known as cavitation.

9.4.1 Elevation of tissue temperature

The mechanical energy of ultrasound can be transformed into heat through absorption in tissue, resulting in the elevation of tissue temperature. Large increases of temperature may cause damage to living cells and pathological changes in tissues. Particularly sensitive to temperature raises is the embryo and foetus **in utero**. Animal experimentation suggests that temperatures exceeding $4^{\circ}C$ above normal body temperature, maintained for long periods of time, may be harmful to the embryo and foetus, leading to a variety of biological changes which can result in abortion, foetal death, or developmental abnormalities, among other effects. Fortunately, the increase in temperature from obstetric ultrasound is well below the thresholds for the initiation of these effects. It requires ultrasonic beam intensities of several hundred mill watts per square centimetre (mW/cm^2) to elevate body temperature by just 1°C. In contrast, the intensity levels used in typical diagnostic instruments fall in the range of a few tens of mW/cm^2. This means that the temperature raise caused by a diagnostic examination will be only a small fraction of 1°C, or insignificant. **The World Federation for Ultrasound in Medicine and Biology (WFUMB) recommends that diagnostic exposures that produce a maximum temperature rise of 1.5°C above the normal physiological level of 37°C may be used without reservation**. From the point of view of thermally-induced effects, the use of diagnostic ultrasound can therefore be considered not only as being safe, but also offering a wide margin of safety in relation to the threshold levels of exposure for deleterious consequences.

9.4.2 Cavitation in liquids

Cavitation is a phenomenon which takes place in liquid cavities exposed to very high intensities of ultrasound. The presence of gas within the volume of the liquid may lead to the formation of micro bubbles during the low pressure phase (rarefaction) of the ultrasound wave cycle. With changes in pressure, the bubbles may grow in size. During the high pressure phase (compression), the micro bubbles may collapse suddenly. Such collapse is associated with the release of large quantities of mechanical energy which can cause biological damage. The peak rarefaction pressure and the peak compression pressure are relevant parameters affecting cavitation. Besides its potential for initiating biological damage through purely mechanical means, cavitation can also be the cause of free radical formation. Free radicals are highly reactive chemical species which can attack bio molecules through a chain of chemical reactions, producing toxic bio products. This leads to damage of biological matter not only locally, but also remotely through diffusion of toxic products.

Experimentation suggests that cavitation is a threshold phenomenon, with threshold intensities increasing as the beam frequency is increased. In the low megahertz frequency range (corresponding to diagnostic frequencies), the thresholds exceed 100 W/cm^2, a thousand times the upper limit of the output intensities for diagnostic equipment. Furthermore, cavitation is unlikely to occur with PW ultrasound when the pulse duration is short, because relatively longer time periods are required for the growth of bubbles.

9.4.3 Other mechanisms.

Besides temperature elevation and cavitation, the deposition of ultrasonic energy in tissue may cause direct mechanical damage to cells and tissues through other mechanisms associated with frictional sheer stress, aggregation of medium particles, and structural changes to particles due to acceleration.

9.5 Assessment of the safety of diagnostic ultrasound.

Ultrasound is a non-ionizing form of energy which has been employed in diagnostic imaging for several decades. Observations of biological changes as a result of exposure to ultrasound have only been made under experimental conditions that differ vastly from those employed in the clinical diagnostic setting. The intensity levels at which adverse effects have been observed are much higher, and the associated exposure durations are generally longer than those typical for diagnostic imaging. **No adverse biological effects have been observed under diagnostic conditions of exposure**. As the pool of scientific knowledge on the safety of ultrasound continues to grow, our assessment of the safety of diagnostic ultrasound must remain under review. From the accumulated scientific knowledge, and with the benefit of several decades of experience, **it can be stated with a high level of confidence that diagnostic ultrasound is a safe method of medical investigation**. Furthermore, it can be stated that **diagnostic levels of exposure are associated with a wide margin of safety in relation to the exposure levels at which adverse effects have been experimentally observed**. Besides the unique diagnostic information provided by ultrasonic imaging vis-à-vis the other imaging modalities, the wide acceptance of ultrasound as a diagnostic tool, and the rapid growth in its usage world-wide, are attributed largely to its safety record.

Even with such positive perception on the safety of ultrasound, experience with ionizing radiation has taught us to approach safety issues concerning exposure to any physical or chemical agent with caution. On the reasonable presumption that the potential risks of exposure increase with increasing energy deposition in tissue, it is appropriate to adopt the strategy of keeping exposures to patients and operators as low as possible consistent with the requirements of the examination. The use of equipment with unnecessarily high output intensities, or any unnecessary prolongation of exposure durations, can be and should be avoided. Efforts to prescribe standards for equipment manufacturers should be encouraged. Operators of ultrasound equipment should avoid the pitfalls of repeated exposures on themselves for purposes of generating demonstration images, a practice that was common among the pioneer users of X-rays, and which later turned out to be most harmful. The mistakes of the past must not be consciously repeated!

Table of content

INTRODUCTION 1
CHAPTER 1 7
   THE NATURE OF ULTRASOUND 7
      1.1 The sound spectrum 7
      1.2 Propagation of ultrasound 7
         1.2.1 Transfer of energy 7
         1.2.2 Pressure waves 8
         1.2.3 Longitudinal propagation 9
         1.2.4. Simple wave parameters. 9
         1.2.5 Velocity of ultrasound 11
CHAPTER 2 15
   GENERATION AND DETECTION OF ULTRASOUND 15
      2.1 The piezoelectric phenomenon 15
      2.2 Production and detection of ultrasound 16
      2.3 Ultrasonic transducers 16
         2.3.1 The single crystal transducer 16
      The crystal element 17
         Example 19
         2.3.1.1 Electrical connections 19
         2.3.1.2 Backing material 19
         2.3.1.3 Acoustic insulator 20
         2.3.1.4 Transducer housing 20
CHAPTER 3 21
   INTERACTION OF ULTRASOUND WITH MATTER 21
      3.1 Acoustic impedance. 21
      3.2 Acoustic boundaries 21
      3.3 Reflection of ultrasound 22
         3.3.1 Specular reflections 22
            3.3.1.1 Implications for diagnostic ultrasound 25
            3.3.1.1.1 Negative role of gas 25
            3.3.1.1.2 Need for coupling gel 26
            3.3.1.1.3 Acoustically homogeneous media 26
            3.3.1.1.4 Transducer matching layer 26
         3.3.2 Scattering of ultrasound (non-specular reflections) 26
      3.4 Refraction of ultrasound 27
      3.5 Absorption of ultrasound 28
      3.6 Beam divergence and interference 30
      3.7 Attenuation of ultrasound in tissues 30
         3.7.1 Ultrasonic half value thickness (HVT) 30
         3.7.2 Acoustic windows and acoustic barriers 31
         3.7.3 Effect of beam frequency 32
CHAPTER 4 33
   INTENSITY OF ULTRASOUND 33
      4.1 Absolute measure of intensity 33
      4.2 Relative intensity 35
         4.2.1 The 3 dB change 35
CHAPTER 5 39
   ULTRASOUND BEAM SHAPE 39
      5.1 General shape of the ultrasound beam 39
      5.2 Factors influencing beam shape 40
         5.2.1 Effect of source size 41
         5.2.2 Effect of beam frequency 42
         5.2.3 Focusing of the ultrasound beam 43
            5.2.3.1 Shape of the crystal element 43

      5.2.3.2 Acoustic lenses..43
      5.2.3.3 Acoustic mirrors..43
      5.2.3.4 Electronic focusing 45
      5.2.3.5 Focus of a transducer, focal zone..45
      5.2.3.6 Classification of focusing 46
  5.3 Optimization of spatial resolution with tissue depth 46
CHAPTER 6 49
  THE ULTRASOUND IMAGE: GENERATION AND DISPLAY 49
  6.1 Basic principle of the ultrasonic image (see Fig. 6.1) 49
  6.2 Electronic processing of signals. 50
  6.3 Echo ranging 51
  6.4 Display modes...53
    6.4.1 The amplitude mode (A-mode)..53
    6.4.2 The brightness mode (B-mode) 54
    6.4.3 The motion mode (M-mode)..55
    6.4.4 Real-time mode 57
    6.4.5 The Doppler mode 57
  6.5 Ultrasound equipment and display modes 59
CHAPTER 7 61
  TRANSDUCERS FOR REAL-TIME IMAGING 61
  7.1 Transducer design 61
    7.1.1 Mechanical transducers..62
    7.1.2 Electronic transducers 64
      7.1.2.1 Geometry of multi crystal arrays 66
      7.1.2.2 Electronic beam focusing...67
    7.1.3 Special applications transducers 68
CHAPTER 8 69
  IMAGE CHARACTERISTICS IN CLINICAL ULTRASOUND..69
  8.1 Spatial resolution 69
    8.1.1 Axial resolution..70
    8.1.2 Lateral resolution 73
      8.1.2.1 Effect of beam width..73
      8.1.2.2 Effect of beam frequency...74
      8.1.2.3 Effect of scan line density..75
  8.2 Contrast resolution 75
  8.3 Temporal resolution 76
  8.4 Limitations due to the velocity of ultrasound 76
  8.5 Optimization of image characteristics 78
    8.5.1 Spatial resolution versus tissue depth 78
      8.5.1.1 Use of multi-frequency transducers...78
    8.5.2 Frame rates, scan line density, and tissue depth. 80
    8.5.3 Summary of compromises 80
      8.5.3.1 For high lateral resolution:...80
      8.5.3.2 For large tissue depths:..80
      8.5.3.3 For very high framing rates (rapid motion) 81
  8.6 Contrast properties of the image display and recording systems. 81
  8.7 Sensitivity of the imaging system 81
  8.8 Artefacts in clinical ultrasound 82
  8.9 Quality control 84
CHAPTER 9 87
  SAFETY OF DIAGNOSTIC ULTRASOUND...87
  9.1 Energy deposition in tissue 87
  9.2 Intensity parameters 88
  9.3 Exposure conditions..88
  9.4 Consequences of high intensity, prolonged exposures 89
    9.4.1 Elevation of tissue temperature..89

- 9.4.2 Cavitation in liquids 90
- 9.4.3 Other mechanisms 91
- 9.5 Assessment of the safety of diagnostic ultrasound. 91

www.ingramcontent.com/pod-product-compliance
Ingram Content Group UK Ltd.
Pitfield, Milton Keynes, MK11 3LW, UK
UKHW051524180426
11947UKWH00018B/1553